Conference Proceedings of the Society for Experimental Mechanics Series

Series Editor

The Conference Proceedings of the Society for Experimental Mechanics Series presents early findings and case studies from a wide range of fundamental and applied work across the broad range of fields that comprise Experimental Mechanics. Series volumes follow the principle tracks or focus topics featured in each of the Society's two annual conferences: IMAC, A Conference and Exposition on Structural Dynamics, and the Society's Annual Conference & Exposition and will address critical areas of interest to researchers and design engineers working in all areas of Structural Dynamics, Solid Mechanics and Materials Research.

Christian Franck • Karen Kasza • Jon Estrada • Rosa De Finis •
Geir Ólafsson • Suhasini Gururaja • Jevan Furmanski • Aaron Forster •
Pavan Kolluru • Mike Prime • Tom Berfield • Cahit Aydiner
Editors

Challenges in Mechanics of Biological Systems and Materials, Thermomechanics and Infrared Imaging, Time Dependent Materials and Residual Stress, Volume 2

Proceedings of the 2023 Annual Conference & Exposition on Experimental and Applied Mechanics

Springer

Editors
Christian Franck
University of Wisconsin-Madison
Madison, WI, USA

Karen Kasza
Columbia University
New York, NY, USA

Jon Estrada
University of Michigan
Ann Arbor, MI, USA

Rosa De Finis
University of Salento
Lecce, Italy

Geir Ólafsson
University of Bristol
Bristol, UK

Suhasini Gururaja
Auburn University
Auburn, AL, USA

Jevan Furmanski
University of Dayton Research Institute
Dayton, OH, USA

Aaron Forster
National Institute of Standards and Technology
Gaithersburg, MD, USA

Pavan Kolluru
Texas A&M University
College Station, TX, USA

Mike Prime
Los Alamos National Laboratory
Los Alamos, NM, USA

Tom Berfield
University of Louisville
Louisville, KY, USA

Cahit Aydiner
Bogazici Universitesi
Istanbul, Türkiye

ISSN 2191-5644 ISSN 2191-5652 (electronic)
Conference Proceedings of the Society for Experimental Mechanics Series
ISBN 978-3-031-50469-3 ISBN 978-3-031-50470-9 (eBook)
https://doi.org/10.1007/978-3-031-50470-9

This Springer imprint is published by the registered company Springer Nature Switzerland AG
The registered company address is: Gewerbestrasse 11, 6330 Cham, Switzerland

Preface

Challenges in Mechanics of Biological Systems and Materials, Thermomechanics and Infrared Imaging, Time Dependent Materials and Residual Stress represents one of five volumes of technical papers presented at the 2023 SEM Annual Conference & Exposition on Experimental and Applied Mechanics organized by the Society for Experimental Mechanics held on June 5–8, 2023. The complete proceedings also include volumes on: Additive and Advanced Manufacturing, Advancement of Optical Methods in Experimental Mechanics, Dynamic Behavior of Materials, Fracture and Fatigue, Inverse Problem Methodologies, Machine Learning and Data Science, and Mechanics of Composite and Multifunctional Materials.

Each collection presents early findings from experimental and computational investigations on an important area within Experimental Mechanics, the Mechanics of Biological Systems and Materials, Thermomechanics and Infrared Imaging, and Time-Dependent Materials being some of these areas.

The Biological Systems and Materials segment of this volume summarizes the exchange of ideas and information among scientists and engineers involved in the research and analysis of how mechanical loads interact with the structure, properties, and function of living organisms and their tissues. The scope includes experimental, imaging, numerical, and mathematical techniques and tools spanning various length and time scales. Establishing this symposium at the Annual Meeting of the Society for Experimental Mechanics provides a venue where state-of-the-art experimental methods can be leveraged in the study of biological and bio-inspired materials, traumatic brain injury, cell mechanics, and biomechanics in general. A major goal of the symposium was for participants to collaborate in the asking of fundamental questions and the development of new techniques to address bio-inspired problems in society, human health, and the natural world. The 2023 Symposium is the 13th International Symposium on the Mechanics of Biological Systems and Materials.

In recent years, the applications of infrared imaging techniques to the mechanics of materials and structures have grown considerably. The expansion is marked by the increased spatial and temporal resolution of the infrared detectors, faster processing times, much greater temperature resolution, and specific image processing. The improved sensitivity and more reliable temperature calibrations of the devices have meant that more accurate data can be obtained than were previously available.

The Time-Dependent Materials track was organized to address constitutive, time (or rate)-dependent constitutive and fracture/failure behavior of a broad range of materials systems, including prominent research in progress in both experimental and applied mechanics. Papers concentrating on both modeling and experimental aspects of Time-Dependent Materials are included.

The Residual Stress track addressed the state of knowledge in the area of experimental techniques for residual stress measurement and experimental characterization of residual stress effects on the performances of materials and engineering systems.

The track organizers thank the presenters, authors, and session chairs for their participation and contribution to these tracks. The support and assistance from the SEM staff are also greatly appreciated.

Madison, WI, USA	Christian Franck
New York, NY, USA	Karen Kasza
Ann Arbor, MI, USA	Jon Estrada
Lecce, Italy	Rosa De Finis
Bristol, UK	Geir Ólafsson
Auburn, AL, USA	Suhasini Gururaja
Dayton, OH, USA	Jevan Furmanski
Gaithersburg, MD, USA	Aaron Forster

College Station, TX, USA
Los Alamos, NM, USA
Louisville, KY, USA
Istanbul, Türkiye

Pavan Kolluru
Mike Prime
Tom Berfield
Cahit Aydiner

Contents

Development of PEEK Matrix Polymer Composite and Additive Manufacturing by Pellet Extrusion Method

Check for updates

Merve Bagci Bilgen and Alaeddin Burak Irez

Abstract Within the context of this study, it is intended to develop new biocompatible polymer-based composite materials. For this purpose, PEEK (polyether ether ketone) was used as the matrix of the composites, and carbon nanotube (CNT) was used as the reinforcement. The PEEK matrix and 1 wt. % CNTs were melt mixed in a twin-screw extruder to obtain the compound in the form of pellets. Then, using these pellets, composites were manufactured using a pellet extruder type 3D printer. After manufacturing, microstructure of the specimens was observed using optical microscopy, and mechanical characterization was performed through three-point bending tests and Charpy impact tests. The most noteworthy result of the microscopy was the absence of any discontinuity between the layers of the specimen for pure PEEK specimens. Furthermore, the mechanical improvement was not apparent by the CNT incorporation for the tested composites. In conclusion, pellet extrusion is thought as a promising tool for the manufacturing of biocompatible materials for biomedical applications.

Keywords Biocompatibility · Pellet extrusion · PEEK · CNT · 3D printing

Introduction

Technological advancements have a significant impact in all areas. In particular, in the manufacturing industry, modern methods are increasingly being used instead of conventional techniques. In this respect, additive manufacturing, also known as 3D printing, is a growing field. A three-dimensional printer is a machine that produces three-dimensional solid objects from a three-dimensional CAD (computer-aided design) file prepared in the digital environment. With three-dimensional printers, models designed in a digital environment can become objects that can be manipulated and examined in a short time [1]. Three-dimensional printer technologies work with the technique of stacking layers on top of each other. However, the methods of creating these layers may differ. For polymers, the most widely known of these methods are the ones that form solid objects by melting the plastic material [2]. The advantages of three-dimensional printers are as follows: the design is easily transferable/shared as it consists of digital data, the ability to make changes and corrections quickly, the ability to easily produce customized products, the efficiency in terms of investment and production, the price of the product can be calculated prior to production, and the use of recyclable materials, which is equivalent to producing a minimum of material waste [3]. There are many types of additive manufacturing methods, and the FDM (fused deposition modeling) additive manufacturing type is one of the most common ones because this technique is simple, cost-effective, and widely available for many different polymers [4]. Studies have been carried out on the FDM technique for a long time, and this method continues to be developed. Despite the advantages mentioned, the FDM technique has the following disadvantages: low extrusion speed and lack of enclosed workspaces to apply this technique to create a large prototype and processing thermoplastic materials such as PEEK (polyether ether ketone) through FDM is a difficult process due to the high melt viscosity. To overcome these challenges, the pellet extrusion technique was used in this project [5]. In pellet extrusion, instead of using the polymer in the filament form with a specific diameter and stiffness, the polymer pellets are fed directly into the extruder of the 3D printer. This offers freedom in the material composition, and it is possible to achieve customized material properties. In this study, PEEK matrix composites were developed to meet the requirements of a wide variety of applications. PEEK is a semi-crystalline thermoplastic used in high-performance engineering applications. Compared to

M. B. Bilgen · A. B. Irez (✉)
Department of Mechanical Engineering, Faculty of Mechanical Engineering, Istanbul Technical University (ITU), Istanbul, Turkey
e-mail: bagcim18@itu.edu.tr; irez@itu.edu.tr

C. Franck et al. (eds.), *Challenges in Mechanics of Biological Systems and Materials, Thermomechanics and Infrared Imaging, Time Dependent Materials and Residual Stress, Volume 2*, Conference Proceedings of the Society for Experimental Mechanics Series,
https://doi.org/10.1007/978-3-031-50470-9_1

other thermoplastics, PEEK has very good thermal stability and excellent resistance to chemicals, solvents, and hot water and can be used continuously for very long periods of time up to 250 °C in air. PEEK also has an important deformation capacity and excellent impact resistance. Besides, the crystalline nature of PEEK provides its high temperature resistance [6]. Similar to many other thermoplastics, the mechanical properties of PEEK can be enhanced by nano-reinforcements such as carbon nanotubes (CNTs) [7]. CNTs exhibit a very high elastic modulus of more than 1 TPa and a strength 10–100 times that of the strongest steel. CNTs also have excellent thermal and electrical properties [8].

Within the context of this study, it is intended to develop new biocompatible polymer-based composite materials that can be utilized in biomedical applications. To this end, PEEK was used as the matrix of the composites, and carbon nanotubes (CNT) were used as the reinforcement. The PEEK matrix and 1 wt% CNT were melt mixed in a twin-screw extruder to obtain the compound in the form of pellets. Then, from these pellets, the composites were fabricated using a pellet extruder type 3D printer. After fabrication, the microstructure of the specimens was observed by optical microscopy, and mechanical characterization was performed by three-point bending tests and Charpy impact tests.

Materials and Methods

In this study, the materials to manufacture the composites were provided by different companies. First, PEEK was supplied from Evonik® under the commercial name VESTAKEEP 2000 G in the form of pellets. It has a density of 1.3 (g/cm^3) and a melting temperature of 340 °C. Next, 0.7 wt% COOH-activated carbon nanotubes (CNT) were provided by Nanografi®, Turkey. The CNT has a density of 2.4 (g/cm^3), a surface area of 65 m^2/g, an outer diameter of 28–48 nm, and a length of 10–25 μm.

After the materials were procured, the composites were manufactured using the additive manufacturing technique of pellet extrusion. Due to the nature of this technique, the composite pellets must be obtained before the specimens are printed. Therefore, the composite pellets were obtained through an extrusion process. Prior to the extrusion process, the PEEK pellets were dried for 3 h in a vacuum oven at 150 °C to remove moisture that may have been trapped due to storage conditions. Then, the composite pellets were fabricated using a Kökbir Makina® co-rotating twin-screw extruder with an L/D ratio of 22, a screw diameter of 12 mm, and a screw speed adjusted to 60 rpm. The twin-screw extruder consists of six zones, and in order to reach the melting temperature of the materials used, the temperature values were set to 47.5 °C for the feed zone, 325 °C for the first zone, 330 °C for the second zone, 340 °C for the third and fourth zones, and 360 °C for the extruder exit. At the exit of the extruder, after the compound has cooled down, the composite is obtained in the form of a filament, and, with the help of a rotary cutter, it is transformed into granules. In the next step, using the pellets from the extrusion process, composites are manufactured using the Yizumi® SpaceA-900-500-S 3D printer. In this 3D printer, the screw diameter was 16 mm, and the screw speed was set to 200 rpm. Regarding the operational parameters of the 3D printer, the extruder temperature was 400 °C, the nozzle diameter was 2.5 mm, the layer thickness was 1.6 mm, and the printed material width was 3 mm. After the composites were fabricated, experimental characterizations were performed.

In the experimental characterizations, the effect of CNT reinforcement as well as the effect of printing direction was examined. The fundamental mechanical properties were identified by three-point bending (3PB) tests. The 3PB tests were performed using a Testometric testing machine according to ASTM D790 test standard, and the crosshead speed was imposed at 10 mm/min. Next, the impact strength of the composites was investigated using a Zwick Roell HIT5P Charpy Universal Tester according to ASTM 6110-10. Each characterization was done using at least five specimens and the standard deviations are given in the results. In addition, specimens from two different composite groups were sectioned and cold mounting was performed. Then, the mounted composites were polished for optical microscopy to observe the microstructure. In addition, scanning electron microscopy (using Tescan Vega 3 SEM) was conducted to observe the damage mechanisms on the fracture surfaces of the failed specimens.

The specimens for the bending tests are given in Fig. 1. The light-colored specimens are pure PEEK specimens, while the black specimens show the CNT-incorporated PEEK matrix composites. In addition, the printing direction can be seen in Fig. 1; Fig. 1b, d shows the longitudinal printing direction, while the others show the transversal printing direction.

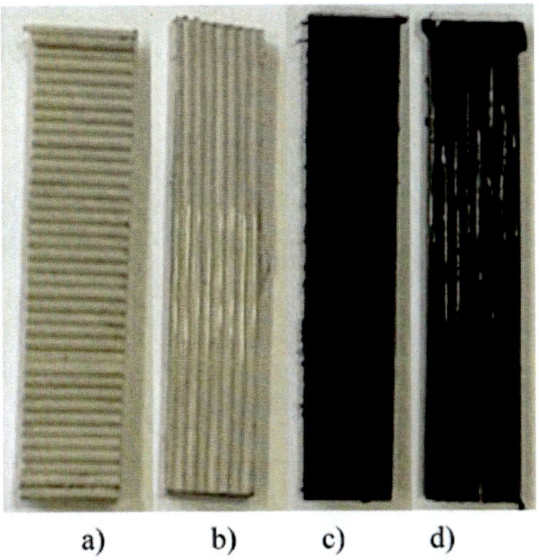

Fig. 1 Three point bending test specimens: (**a**) PEEK transversal, (**b**) PEEK longitudinal, (**c**) PEEK/CNT transversal, (**d**) PEEK/CNT longitudinal

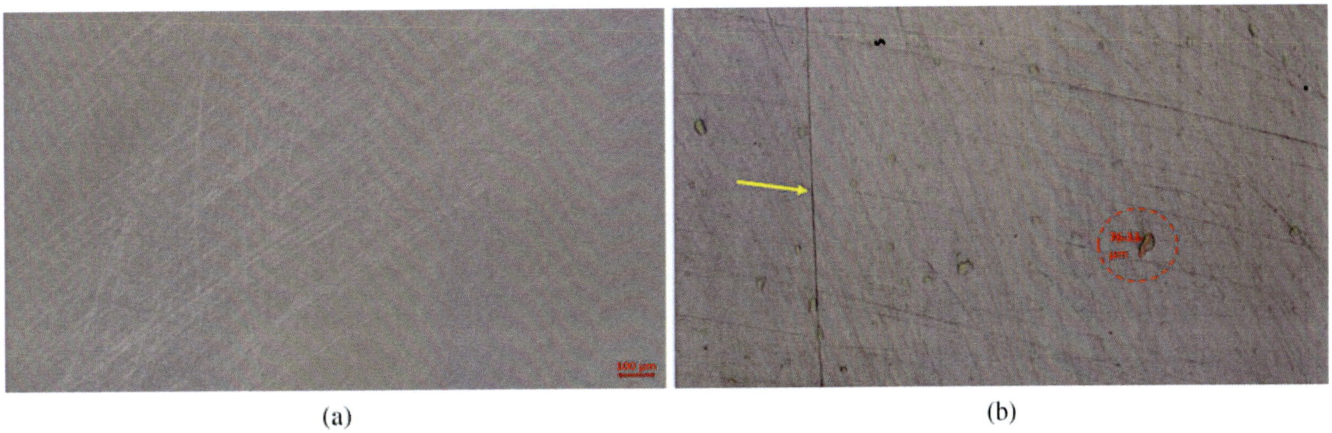

Fig. 2 Optical microscope images: (**a**) PEEK specimen, (**b**) PEEK/CNT composite

Results and Discussion

After the manufacturing steps, sections of the printed specimens were taken and then mounted into resin. After polishing the specimens, optical microscopy was performed to observe the microstructure of the composites. In some 3D printing techniques, voids and fusion problems between successive layers can be observed in the printed part, resulting in degradation of mechanical properties. On the other hand, as can be seen in Fig. 2a, there are some scratches due to polishing in the sectioned surface. Nevertheless, the extrusion of PEEK pellets does not have discontinuities, which is an advantage for this technique. After incorporation of CNTs, some manufacturing-related problems are observed as shown in Fig. 2b. In Fig. 2b, as indicated with a yellow arrow, a line between the layers is observed. Besides, inside the dashed circles, some agglomerated CNTs are observed.

After the observation of the microstructures, the mechanical properties of the printed part were studied by three-point bending tests. Additive manufacturing techniques can cause anisotropy in the manufactured specimens with respect to the feed direction of the printhead. Therefore, this effect must also be tested after fabrication. In this regard, specimens printed parallel to the printhead feed direction are referred to as longitudinal specimens, and specimens with an orientation perpendicular to the printhead feed direction are referred to as transverse specimens. After performing the tests, the flexural moduli as a function of the tested parameters are given in Fig. 3. According to Fig. 3, CNT incorporation reduced the stiffness of the composites for both printing directions. This may be associated with the agglomeration of CNTs in the printed specimens due to the intense van der Waals interactions between the carbon atoms and some manufacturing deficiencies [9].

In addition, the calculated flexural modulus in the pure PEEK specimens with transverse printing direction was found to be 15% higher than the values in the samples with longitudinal printing direction. The printing direction also affected the strength of the pure PEEK material as given in Fig. 4.

From Fig. 4, for pure PEEK specimens, the effect of printing direction is not significant. As shown in the microscopic images, the interfaces between successive layers are defect-free. Therefore, this may be the reason for the unchanged strength.

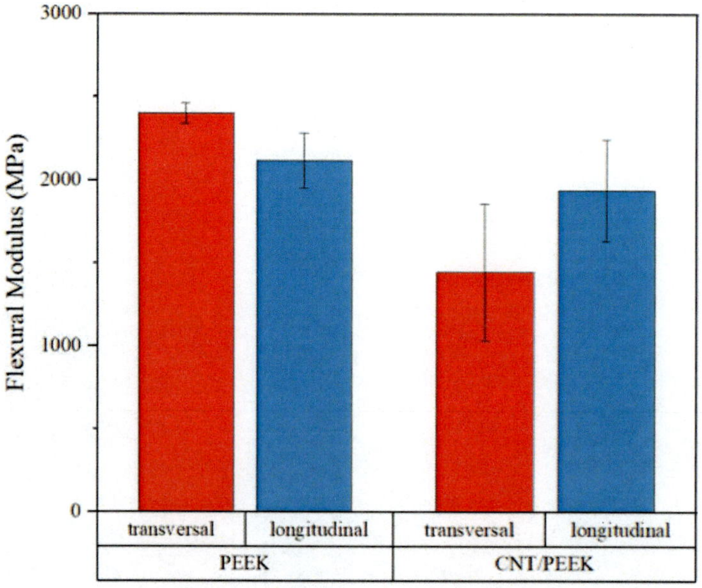

Fig. 3 Flexural moduli of the printed specimens with respect to CNT incorporation and printing direction

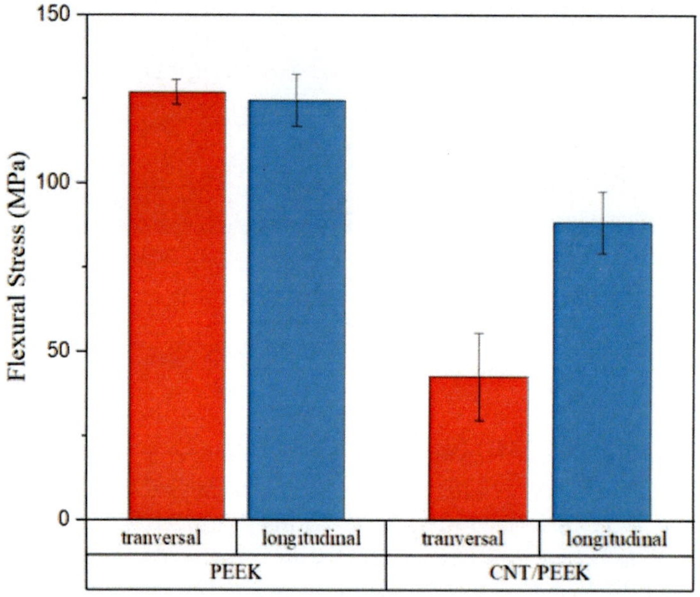

Fig. 4 Maximum flexural stress of the printed specimens with respect to CNT incorporation and printing direction

On the other hand, CNT incorporation again lowered the strength similar with stiffness. This reduction is more apparent for the transversally printed specimens. In addition, the standard deviation of the strength of CNT-reinforced specimens is larger than that of pure PEEK specimens. This may be the result of the inhomogeneity of CNT distribution. Finally, the elongation capacity of the printed parts was examined from bending tests and is given in Fig. 5. For pure PEEK specimens, it is clearly shown in Fig. 5 that longitudinally printed pure PEEK specimens can deform to failure by twice as much as transversely printed pure PEEK specimens. On the other hand, the decreasing trend of mechanical properties with the incorporation of CNTs did not change for the elongation at break. Discontinuities and inhomogeneities can cause stress concentration in the microstructure, leading to premature failure. In addition, longitudinally printed specimens may deform more than transversely printed parts. This is an expected result since transversely printed specimens have more layers.

After the three-point bending test, Charpy impact tests were carried out and results are given in Fig. 6. Upon examination of the results, it was found that the Charpy impact strength of the PEEK samples in the longitudinal printing direction was higher than the other samples. It was observed that the lowest impact strength was found in the transversal printing direction of the PEEK/CNT samples. It was expected that, due to the damping characteristics of CNTs, the impact resistance might

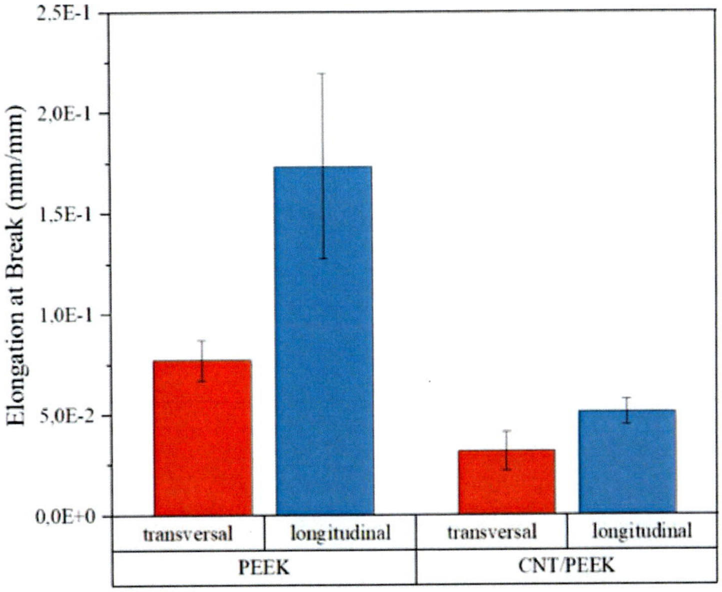

Fig. 5 Elongation at break of the printed specimens with respect to CNT incorporation and printing direction

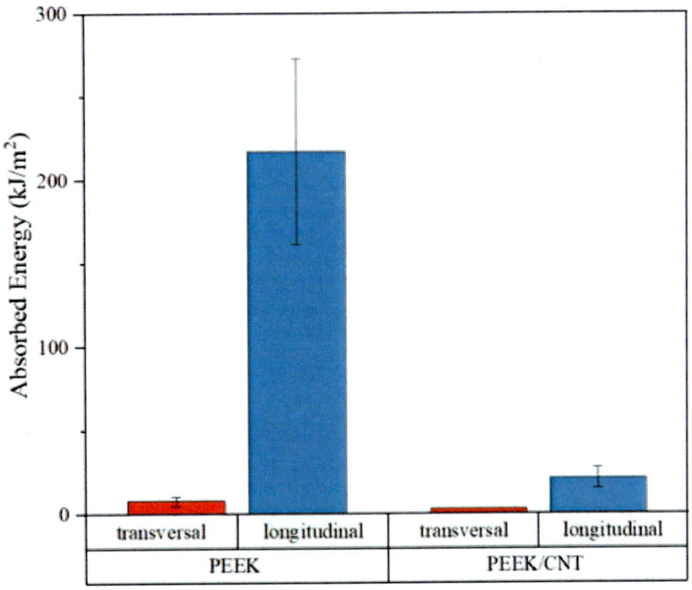

Fig. 6 Charpy impact test results with respect to CNT incorporation and printing direction

have been improved. However, as with the bending tests, no improvement was observed in the impact resistance of the composites. This may be due to manufacturing issues.

After completing the mechanical characterizations, SEM fractography was performed on the failed specimens. In Fig. 7, the fracture surface of the pure PEEK specimen is given, and a smooth fracture surface is considered an indicator of ductile behavior. In addition, no discontinuities were observed in the microstructure, which is consistent with the optical microscopy.

Conclusion

In this study, a composite material was developed for use in biomedical applications. The novelty of this study is the manufacturing of this composite by the additive manufacturing technique of pellet extrusion. As reinforcing agents, CNTs were incorporated into the PEEK matrix by twin-screw extrusion. Then, the composite in pellet form was used by the 3D printer to fabricate the specimens for mechanical characterizations. From microstructural observations, no discontinuity was

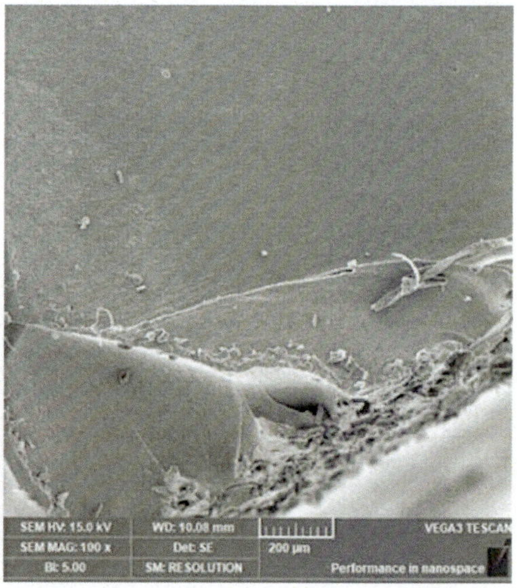

Fig. 7 SEM fractography of pure PEEK specimen

observed for the pure PEEK specimens. However, the CNT addition resulted in some homogeneity issues. This is probably associated with the high van der Waals forces between the carbon atoms and the π-π interactions. For these reasons, the expected improvement in mechanical properties was not observed. In addition, the effect of the printing direction was also investigated, and anisotropy in the specimens with respect to the printing direction was observed. These results were obtained on a laboratory scale, and it is expected that eliminating the deviations by using more advanced manufacturing facilities, these composites can be promising in the development of biomedical implants.

Acknowledgments The authors thank Dr. Philip Engel from Evonik Industries, Germany Medical Systems Branch, for material support and Mr. Erdem Ozdemir from Yizumi GmbH for the additive manufacturing support. This research was supported by the Istanbul Technical University Office of Scientific Research Projects (ITU BAPSIS), under grant FHD-2022-44279.

References

1. Prince, J.D.: 3D printing: an industrial revolution. J. Electron. Resour. Med. Libr. **11**(1), 39–45 (2014). https://doi.org/10.1080/15424065.2014.877247
2. Kruth, J.P., Leu, M.C., Nakagawa, T.: Progress in additive manufacturing and rapid prototyping. Cirp Ann. **47**(2), 525–540 (1998)
3. Şahin, K., Bülent Onur, T.: ÜÇ BOYUTLU YAZICI TEKNOLOJİLERİNİN KARŞILAŞTIRMALI ANALİZİ. Stratejik ve Sosyal Araştırmalar Dergisi. **2**(2), 97–116 (2018)
4. Cuiffo, M.A., Snyder, J., Elliott, A.M., Romero, N., Kannan, S., Halada, G.P.: Impact of the fused deposition (FDM) printing process on polylactic acid (PLA) chemistry and structure. Appl. Sci. (Switzerland). **7**(6) (2017). https://doi.org/10.3390/app7060579
5. Wang, Z., Liu, R., Sparks, T., Liou, F.: Large-scale deposition system by an Industrial Robot (I): design of fused pellet modeling system and extrusion process analysis. 3D Print Addit. Manuf. **3**(1), 39–47 (2016). https://doi.org/10.1089/3dp.2015.0029
6. Thiruchitrambalam, M., Bubesh Kumar, D., Shanmugam, D., Jawaid, M.: A review on PEEK composites – manufacturing methods, properties and applications. Mater. Today Proc. **33**, 1085–1092 (2020). https://doi.org/10.1016/j.matpr.2020.07.124
7. Arif, M.F., Kumar, S., Varadarajan, K.M., Cantwell, W.J.: Performance of biocompatible PEEK processed by fused deposition additive manufacturing. Mater. Des. **146**, 249–259 (2018). https://doi.org/10.1016/j.matdes.2018.03.015
8. Norizan, M.N., et al.: Carbon nanotubes: functionalisation and their application in chemical sensors. RSC Adv. **10**(71), 43704–43732 (2020). https://doi.org/10.1039/d0ra09438b
9. Irez, A.B., Bayraktar, E., Miskioglu, I.: Toughening mechanism in epoxy resin modified recycled rubber based composites reinforced with gamma-alumina, graphene and CNT. In: Mechanics of Composite, Hybrid and Multifunctional Materials, Volume 5: Proceedings of the 2018 Annual Conference on Experimental and Applied Mechanics. Springer International Publishing (2019)

Parkinsonian Hand Tremor Mitigation with Enhanced Wearable Device

Zach B. Gerards, J. Gus Stucki, and Timothy A. Doughty

Abstract Parkinson's disease is a neurodegenerative disease affecting over eight million people globally. Among its symptoms is an involuntary rhythmic tremor in the body that manifests in the hands. These hand tremors affect one's ability to perform basic motor functions, often leading to frustration, social anxiety, and isolation. Approaches to mitigating these tremors generally involve surgery or medications. Those approaches that are non-invasive are often still expensive or conspicuous. The goal of this research is to make a wearable, non-invasive, mechanical device that can significantly diminish hand tremors. The device is targeted to be accessible to large populations through low cost, ease of use and care, and adjustability for individual needs. The research builds from an ongoing study where a simplified proof of concept was established that reduced tremors in some instances by over 70%. In this study, novel kinematic mechanisms have been introduced to optimize the range of effectiveness while reducing the size and weight of the device by more than 20%. Prototyping and testing have been performed on a mechanical test hand, and initial testing shows quantifiable improvements in size, weight, and effectiveness.

Keywords Frequency · Parkinsonian tremor · Essential tremor · Tremor suppression · Assistive technology

Introduction

Parkinson's disease (PD) is a progressive neurological disorder. It is caused by the gradual deterioration of neurons in the brain. Many of the symptoms are due to a loss of neurons that produce dopamine, which causes atypical brain activity, leading to impaired movement and other symptoms of Parkinson's disease [1]. Approximately 500,000 Americans are diagnosed with PD, but given that many individuals go undiagnosed or are misdiagnosed, the actual number is likely much higher. Some experts estimate that as many as one million Americans are affected [2]. Young adults rarely experience Parkinson's disease. It ordinarily begins in middle or late life, and the risk increases with age. People usually develop the disease around age 60 or older [1].

Symptoms of Parkinson's disease can be physical and psychological, ranging from the well-known muscle tremors and impaired movement to speech issues and cognitive difficulties. These symptoms can result in major quality of life problems, such as poor social functioning, emotional health issues, poor nutrition, and poor hygiene [3]. The basic daily tasks that people take for granted can be much more difficult and strenuous for someone suffering with PD. Tremors and muscle fatigue can make fine motor control difficult in many circumstances. The most common type of tremor is the "pill rolling" tremor, where it appears a person is repeatedly rolling a pill or small object between their thumb and index finger [4].

There is currently no cure for Parkinson's disease, but treatments are available to help relieve the symptoms and maintain quality of life. These treatments include supportive therapies, medication, and surgery. The most common medication is a combination of carbidopa and levodopa, a medication that is converted into dopamine in the brain [5]. This medication, along with other similar ones, can have side effects including dizziness, mouth and throat pain, confusion, weakness, and potentially more serious ones like hallucination, fever, vomiting, and uncontrolled movements [6]. The effectiveness of these medications also wanes over time. The surgical options are often invasive and involve risks to the patient. Deep brain

Z. B. Gerards · J. G. Stucki · T. A. Doughty (✉)
Shiley School of Engineering, University of Portland, Portland, OR, USA
e-mail: gerards24@up.edu; stucki24@up.edu; doughty@up.edu

C. Franck et al. (eds.), *Challenges in Mechanics of Biological Systems and Materials, Thermomechanics and Infrared Imaging, Time Dependent Materials and Residual Stress, Volume 2*, Conference Proceedings of the Society for Experimental Mechanics Series,
https://doi.org/10.1007/978-3-031-50470-9_2

stimulation may be offered to those with advanced PD, but it involves an invasive procedure where electrodes are implanted on specific parts of the brain and are controlled by an exterior device that sends electrical impulses [5].

Given all of the alternatives, a device that is non-invasive, affordable, and easy to use may be an attractive approach to suppressing hand tremors and returning some quality of life back to a patient. There are some alternatives on the market, and they primarily use a range of suppression methods: active, semi-active, and passive. Active suppression generates an active force or motion that counteracts the tremor, whereas semi-active and passive suppression utilizes methods of energy dissipation or absorption [7]. These devices tend to be expensive or bulky—downsides that may discourage someone from using them. One product available on the market has a price of $749 [8], and some of the wearable tremor suppressing devices can weigh more than half a kilogram [9]. The research presented here leverages its merit in contrast to these costs.

Background

The pursuit of non-invasive tremor-reducing devices has been underway for several years. This project builds on the foundations set by T. Doughty and N. Bankus in establishing a model of the human hand joint as a single degree of freedom mass-spring-damper system and upon multiple other studies stemming from that publication. Researchers in this series of projects have created multiple proof of concept designs for Parkinsonian tremor suppression. The internal mechanisms of these devices can be effectively modelled as a nonlinear mass-spring-damper system, which when applied to the hand is able to significantly reduce the amplitude of the Parkinsonian tremor. These prototypes have been developed in the form of eating utensils, drinking glasses, and several wearable devices [10–12].

These are excellent examples of the potential effectiveness of such a device. They are non-invasive and purely mechanical, making for a simple and affordable improvement to quality of life for those afflicted with tremors. While the work on the utensil and cup is concluded, iterations are still ongoing for the wearable device, with a goal to optimize the ergonomics and weight. A wearable device should be unobtrusive as it is intended for users with mobility issues. Furthermore, most of the designs have been unable to suppress a sufficiently wide range of tremors. Those living with PD may live with tremors with frequencies of 4–6 Hz [10], while those with other tremor-inducing conditions could experience any from 3–10 Hz [13]. To help guide performance goals for a release candidate design, Table 1 below quantifies the design criteria set by S. Winston, R. Dehmer, and T. Doughty to guide performance and physical characteristics of a successful device design.

Two wearable devices documented in the paper by Winston, Dehmer, Horen, and Doughty, titled "The Pin" and "The String," demonstrated the ability to suppress a wide spectrum of tremor frequencies. Both devices notably featured new mechanisms to adjust the range of tremor suppression, with the String demonstrating an impressive optimal range of 4.7–11.5 Hz. However, the adjustment systems on both prototypes were imprecise and would be difficult to use for those suffering from tremors. Additionally, they were still not close to an acceptable level of size and weight—the Pin being larger than desired and the String being heavier.

The purpose of this study is to build on the design principles of the Pin and the String by introducing novel concepts to better meet the criteria above. Two notable changes are present in this prototype. The first is a new adjustment mechanism through the implementation of two orthogonal tremor suppression systems in the same housing. By rotating a dial on the

Table 1 Design criteria

Criteria	Weight	Description
Effectiveness	Essential (5)	The device is effective at suppressing tremors when compared to other solutions. It is effective for frequencies ranging from 4 to 12 Hz
Adjustability	Essential (5)	Every Parkinson's patient is different. The device is meant to help as many people as possible, so the device is designed such that the user can adjust two key features: the effective frequency of the device and the fit of the device on their hand
Health and safety	Essential (5)	The device is safe for all users and non-invasive
Mass	Important (4)	The device is under 450 g
Aesthetics	Important (4)	The user is confident while wearing the device. It does not draw attention to itself while still being aesthetically pleasing. The device should be small to help attain this goal
Ergonomics	Prominent (3)	The device fits and feels comfortable on the user's hand
Cost	Desired (2)	The cost of the device is competitive, relative to other solutions
Durable and sustainable	Optimal (1)	The device is designed to last a long time and through many climates. Parts should be easily replaceable if something breaks. Materials are environmentally friendly and recyclable

device, the degree to which each system is engaged varies. This creates an analog level of adjustment for the device's effective stiffness that relates to the optimized frequency of tremor suppression. The second aspect introduces a novel design geometry, hereafter known as "Mechanism 2," that increases the device's inertia without increasing its total mass. This allows for the suppression of lower frequency tremors without sacrificing the mass design criteria of Table 1.

Methods

A mechanical hand assembly was constructed to test each device in a controlled and repeatable environment. The foundation of this project is the work by T. Doughty and N. Bankus in establishing a model of the human hand and wrist joint roughly similar to a mass-spring-damper system. This testing system is used to simulate the behavior of the human hand. The mass, stiffness, and damping can be adjusted to represent values similar to those estimated in the behavior of a typical human hand. The mass used for the hand was 400 g, and the stiffness and damping can be adjusted using the slotted design of the attachment to the rotating base. Once in place, these parameters were verified through a measurement of the mechanical hand's response to known stimuli. In this case, the bearings and mass of the hand were determined to have an effective damping ratio comparable to that of a human hand during knocking tremors at 0.038. This was set according to previous research which describes the natural frequency of a human wrist joint [10] (Fig. 1).

Tests were performed on this mechanism by recording the system performance over a range of sinusoidal inputs generated by a dynamic shaker, simulating a range of frequencies of hand tremor. Stimulus and response accelerations were logged using 2D accelerometers mounted on the shaker arm and hand, respectively. After these signals passed through a low-pass filter set at 16 Hz, they were recorded, and the shaker progressed to the next discrete tremor frequency. LabVIEW programming then assembled this data into a frequency response function (FRF), from which the magnitude of tremors experienced at each frequency can be interpreted. Figure 2 illustrates this process.

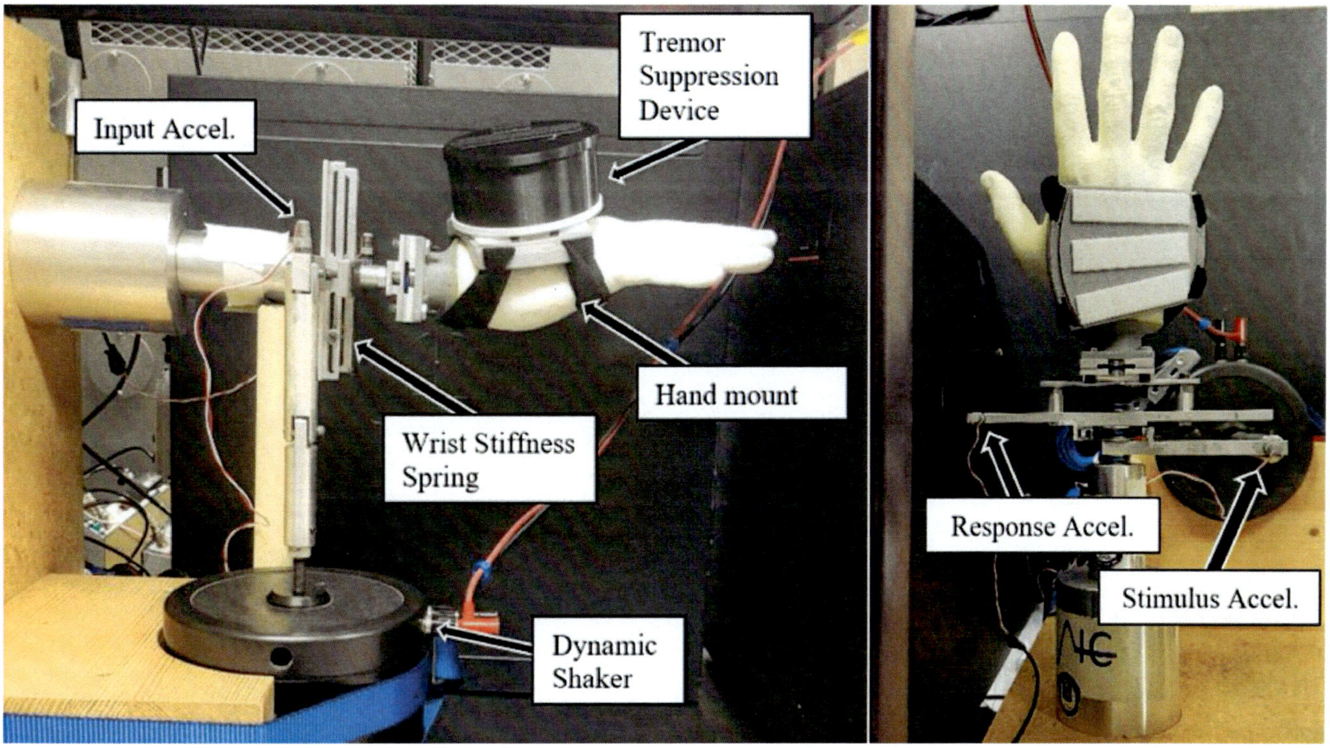

Fig. 1 Annotated view of testing mechanism with tremor suppressor attached. (**a**) Side view of testing mechanism. (**b**) Top view of testing mechanism

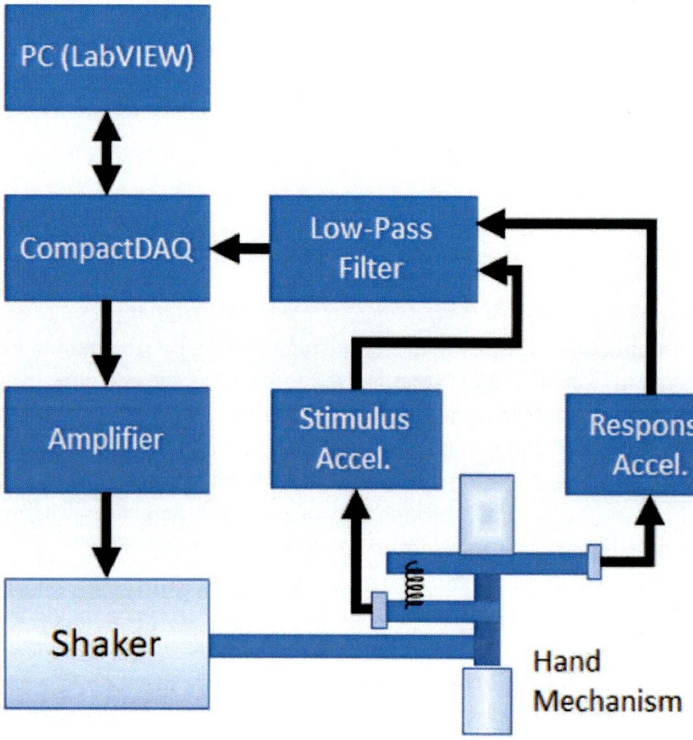

Fig. 2 Block diagram of the test environment

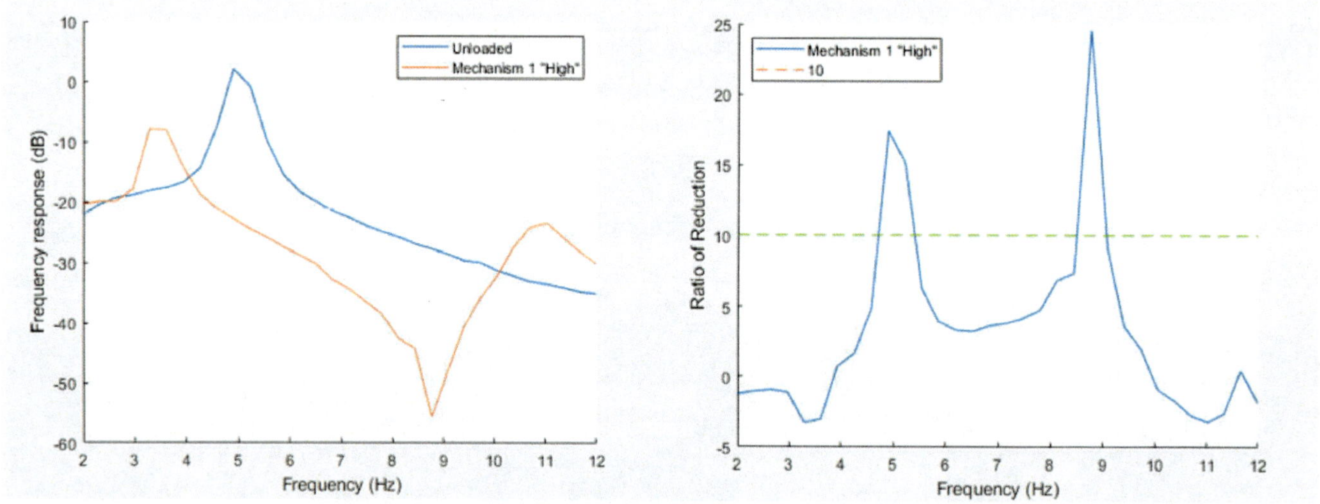

Fig. 3 (**a**) Device using mechanism 1 and the "High" adjustment. (**b**) Associated ratio of reduction

Results

The following results were generated via MATLAB by overlaying the frequency response of the mechanical hand with and without the device. All tests were run from 2 to 12 Hz in 32–40 even steps, depending on the necessary fidelity to obtain clear results. The device was tested for each of its configurations and for a comparison between its movement mechanisms. The lowest point on the frequency response curve dictates the optimally suppressed frequency at that configuration. The ratio of tremor reduction can be solved by comparing the amplitude of the frequency response in dB of the bare mechanical hand to the response of the hand and device at the same frequency. The effective range was determined to be the range of frequencies for which the device had a ratio of reduction greater than 10, corresponding to a measured amplitude 10 times less than that of the hand without the device (Figs. 3, 4, and 5).

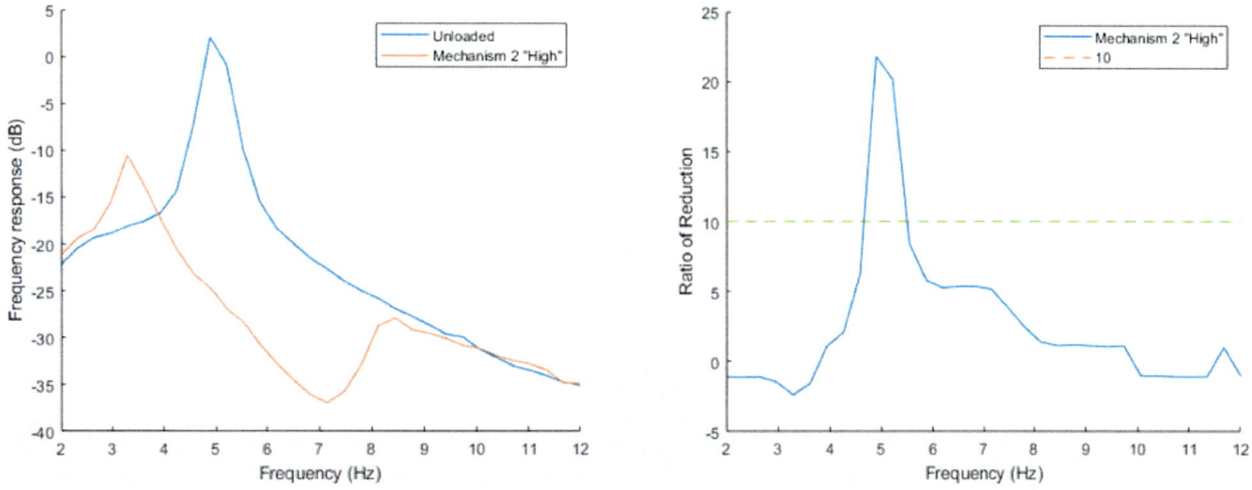

Fig. 4 (**a**) Device using mechanism 2 and the "High" adjustment. (**b**) Associated ratio of reduction

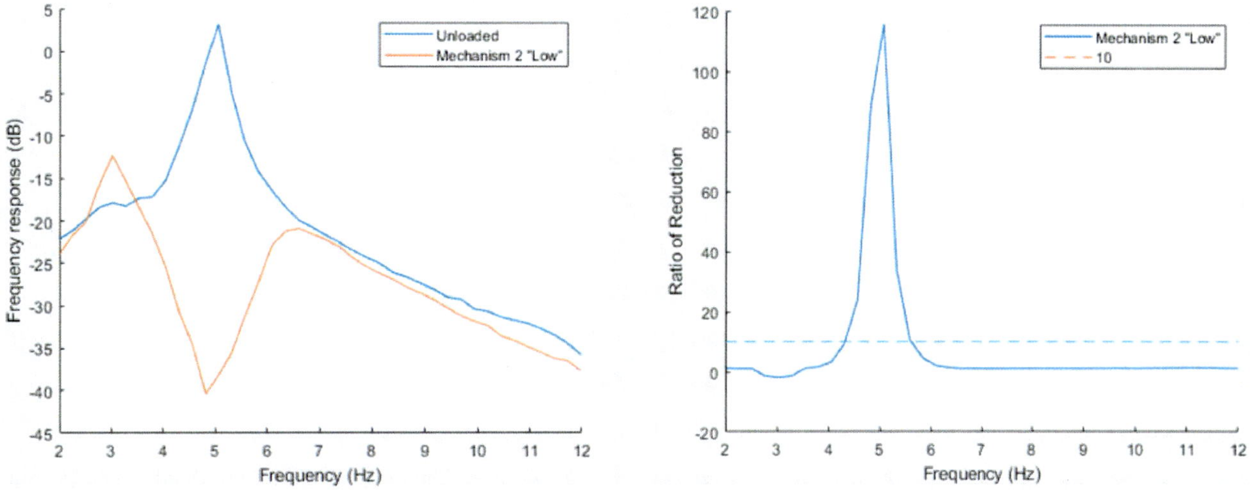

Fig. 5 (**a**) Device using Mechanism 2 and the "Low" adjustment. (**b**) Associated ratio of reduction

Table 2 Frequency response data of several settings of the tremor suppressor device prototype

Mechanism and adjustments	Optimal suppression frequency [Hz]	Ratio of reduction at lowest response	Effective range(s) [Hz]
Mechanism 1 "High" setting	8.8	17.4, 24.6	4.7–5.4, 8.5–9.1
Mechanism 2 "High" setting	7.2	21.8	4.7–5.5
Mechanism 2 "Medium" setting	6.5	38.8	4.6–6.1
Mechanism 2 "Low" setting	4.8	115.6	4.3–5.6

Discussion

While utilizing Mechanism 1, the device was effective at reducing hand tremors from 4.1 to 10.1 Hz with optimum performance at two effective ranges, one at 4.7–5.4 Hz, and the other at 8.5–9.1 Hz (Table 2). This is because the device negates the spike at the 5 Hz response of the unloaded hand, while its optimal suppression frequency is at 8.8 Hz. When using Mechanism 2 and the same "High" adjustment setting, the optimal frequency drops to 7.2 Hz. The effective range also slightly increases to 4.7–5.5 Hz. Adjusting the device to the "Medium" setting for Mechanism 2 displays an optimal frequency of 6.5 Hz and a higher peak ratio of reduction of 38.8. The effective range is slightly wider still, at 2.6–6.1 Hz. Finally, at the "Low" setting for Mechanism 2, we see the most efficacy. The optimal frequency is 4.8 Hz, and a ratio of reduction of 115.6 is observed. The effective range is 4.3–5.6 Hz.

It can be expected that the ratio of reduction will be the highest when the optimal suppression frequency more closely aligns with the natural frequency of the unloaded hand mechanism, as it is a comparison of amplitudes. The device in all configurations has a relatively small optimum frequency range, but further designs may work to improve this result. Given the effective frequencies observed at each adjustment setting, there is a clear decreasing trend from "High" to "Low" and from Mechanism 1 to Mechanism 2.

These results show that we can decrease the effective frequency and increase the effective mass simply by changing the mechanism of movement, potentially allowing for future modifications of the device to become much more effective without becoming less ergonomic. We can experience the benefits of having a larger mass in a smaller physical space. The optimal frequency can be further changed among the same mechanism by moving to different adjustment settings, which is a very simple and quick process—one that could allow someone suffering from PD to easily accommodate changes in the frequency of their tremors.

The device weighs almost exactly the same for both mechanisms, at about 180 g, meeting the mass criterion previously established (Table 1), and both are constructed out of PLA and ABS plastics, rubber, acrylic, and steel. The device is housed in a bulky shell, failing the ergonomics and aesthetics criterion, but the housing is mostly used to obfuscate the internal mechanisms of the device. Future improvements will certainly reduce the size to an acceptable level. The device is safe and non-invasive, as it can simply be mounted on the back of the hand and easily removed, meeting the health and safety criterion. The mounting mechanism as it stands has some padding but is not very comfortable over extended periods of time. The cost of the materials required is very low relative to other solutions, so the cost criteria is met, but the cost is expected to increase somewhat as higher fidelity iterations are developed. The long-term durability remains untested, but the device successfully endured many hours of rigorous shaking during testing over the duration of the project.

Conclusion

This device remains a proof of concept like previous prototypes. It is among the lightest, but is not the smallest made so far in this family of studies. That said, it offers promising new solutions to the reduction of both size and mass, making important steps toward a lightweight and ergonomic release candidate design. The cost of production remains low at this point, which will allow for wide accessibility in today's market.

Efficacy has been demonstrated in the new adjustment system, and while it currently does not cover the same effective range as the previous String design at 4.8–7.2 Hz versus 4.7–11.5 Hz, it allows for more precise adjustment and better ease of use. The system also shows the promise of scalability, and future iterations of this design are likely to continue to widen its effective range. Actuation Mechanism 2 has also proven successful as a method of increasing the effective mass of the system. This allows for a significant increase in performance per mass of the system in comparison with the more conventional Mechanism 1. Further improvements to these systems are currently under study.

It must be noted also that while the "Low" setting on Mechanism 2 has the greatest ratio of reduction of the three settings tested in this trial, at 115.6 versus 21.8 and 38.8 for the "High" and "Medium" settings, respectively, this is primarily due to the unloaded hand's tremor frequency remaining static at 5 Hz for all tests. For tremors at frequencies closer to the optimal frequencies at those settings, the ratio of reduction would become much higher. Future work in this project will analyze this.

One consideration for further work in this project is to continue the validation of the device through criteria based on human subject testing. Many of these design goals are arbitrarily chosen and could be subject to change upon end-user input and testing.

Acknowledgments Thanks to the Shiley School of Engineering for its essential funding of this research. Additional thanks to the team's advisor, Dr. Timothy Doughty, for his instruction, counsel, and insight on these assistive devices.

References

1. Parkinson's disease – symptoms and causes. Mayo Foundation for Medical Education and Research, [Online]. Available: https://www.mayoclinic.org/diseases-conditions/parkinsons-disease/symptoms-causes/syc-20376055
2. Parkinson's disease: challenges, progress, and promise. National Institute of Neurological Disorders and Stroke, [Online]. Available: https://www.ninds.nih.gov/current-research/focus-disorders/focus-parkinsons-disease-research/parkinsons-disease-challenges-progress-and-promise
3. Chiong-Rivero, H., Ryan, G.W., Flippen, C., et al.: Patients' and caregivers' experiences of the impact of Parkinson's disease on health status. PubMed Central. (2011). https://doi.org/10.2147/PROM.S15986

4. Geng, C.: Pill-rolling tremor: causes, treatment, and more. MedicalNewsToday, [Online]. Available: https://www.medicalnewstoday.com/articles/pill-rolling-tremor
5. Parkinson's disease – diagnosis and treatment. Mayo Foundation for Medical Education and Research, [Online]. Available: https://www.mayoclinic.org/diseases-conditions/parkinsons-disease/diagnosis-treatment/drc-20376062
6. Levodopa and carbidopa: MedlinePlus drug information. U.S. National Library of Medicine, [Online]. Available: https://medlineplus.gov/druginfo/meds/a601068.html
7. Mo, J., Priefer, R.: Medical devices for tremor suppression: current status and future directions. PubMed Central. (2021). https://doi.org/10.3390/bios11040099
8. Tremelo, an essential tremor device for hand and arms. Five Microns, [Online]. Available: https://fivemicrons.com/products/tremelo/
9. Nguyen, H.S., Luu, T.P.: Tremor-suppression orthoses for the upper limb: current developments and future challenges. Frontiers, [Online]. Available: https://www.frontiersin.org/articles/10.3389/fnhum.2021.622535/full
10. Doughty, T., Bankus, N.: Mechanical modeling and design for reduction of Parkinsonian hand tremor. In: American Society of Mechanical Engineers, vol. 2, pp. 521–527 (2010)
11. Winston, S.E., Dehmer, R.C., Doughty, T.A.: Parkinsons disease: tremor suppression with wearable device. Volume 5: Biomedical and Biotechnology (2021). https://doi.org/10.1115/IMECE2021-70910
12. Zach, H., Dirkx, M., Bloem, B.R., Helmich, R.C.: The clinical evaluation of Parkinson's tremor. J. Parkinson's Dis. **5**(3), 471–474 (2015)
13. Charles, P.D., Esper, G.J., Davis, T.L., Maciunas, R.J., Robertson, D.: Classification of tremor and update on treatment. Am. Fam. Physician. **59**, 1565–1572 (1999)

Quantitative Measurement of Viscoelastic Properties of Soft Membranes Subjected to Finite Deformations Based on Optical Coherence Elastography

O. Balogun and Z. Wang

Abstract Glaucoma is a leading cause of irreversible blindness that affects over 60 million people worldwide. Glaucomatous eyes are associated with risk factors such as elevated intraocular pressure (IOP) and low corneal hysteresis. Reliable non-invasive measurement of IOP remains a formidable challenge that limits the accurate diagnosis of glaucoma and associated intervention therapies. This work investigates the propagation of shear-dominated elastic waves in hydrostatically inflated corneal tissue phantoms based on the optical coherence elastography (OCE) technique. Unlike previous approaches reported in the literature, we analyze the dispersion relation of guided elastic waves in the phantoms by accounting for both small amplitude viscoelastic wave propagation and finite static deformations. The analytical approach we adopted will enable the determination of the storage and loss shear moduli dependence on finite strains in the cornea that results from hydrostatic pressures. This work provides a modeling and experimental framework for accurately characterizing viscoelastic properties and the IOP of corneal tissues.

Keywords OCE · Shear wave propagation · Hyper-viscoelasticity

Introduction

Glaucomas are a group of optic neuropathies that lead to vision loss and blindness [1]. Elevated intraocular pressure (IOP) is a significant risk factor for the development of glaucoma, and IOP reduction is the only treatment shown to be effective in stopping or slowing the disease progression [2–11]. Recent studies have shown that current clinical methods for the measurements of the IOP can be inaccurate in some patients [11–14] and lead to erroneously low or high IOP measurements. Currently utilized measurement techniques do not account for variations in the corneal structure and biomechanical properties. Thus, there is an unmet clinical need for improved methods to accurately measure the IOP and to account for the intrinsic variability of the cornea biomechanical properties.

Corneal tissue is a complex, layered material that exhibits elastic, viscous, and anisotropic deformation in response to mechanical stimuli. The cornea is subjected to a biaxial tensile stress state at elevated IOP levels due to fluid loading from the anterior chamber. The resulting cornea deformation depends on its five-layered microstructure, the stroma being the dominant layer. Recently, the optical coherence tomography (OCT) technique emerged as a powerful technique routinely used in ophthalmology practice to characterize cornea tissue morphology with micron-scale lateral spatial resolution. Integrating dynamic mechanical stimuli with the OCT allows for local excitation, detection of elastic waves [15], and measurement of cornea tissue elastic properties. The latter approach is commonly called the optical coherence elastography (OCE) technique. Recently, the OCE technique was used to characterize the IOP-dependent elastic wave speeds and the anisotropic shear stiffness constants in porcine corneal tissue [16, 17]. However, while these measurements are promising for OCT/OCE diagnosis of glaucomas, they ignore the influence of large static deformations of the cornea that occur at elevated IOP levels. The large deformation can significantly modulate the viscoelastic properties due to the realignment of collagen fibrils in the

O. Balogun (✉)
Department of Mechanical Engineering, Northwestern University, Evanston, IL, USA

Department of Civil and Environmental Engineering, Northwestern University, Evanston, IL, USA
e-mail: o-balogun@northwestern.edu

Z. Wang
Department of Civil and Environmental Engineering, Northwestern University, Evanston, IL, USA

C. Franck et al. (eds.), *Challenges in Mechanics of Biological Systems and Materials, Thermomechanics and Infrared Imaging, Time Dependent Materials and Residual Stress, Volume 2*, Conference Proceedings of the Society for Experimental Mechanics Series,
https://doi.org/10.1007/978-3-031-50470-9_3

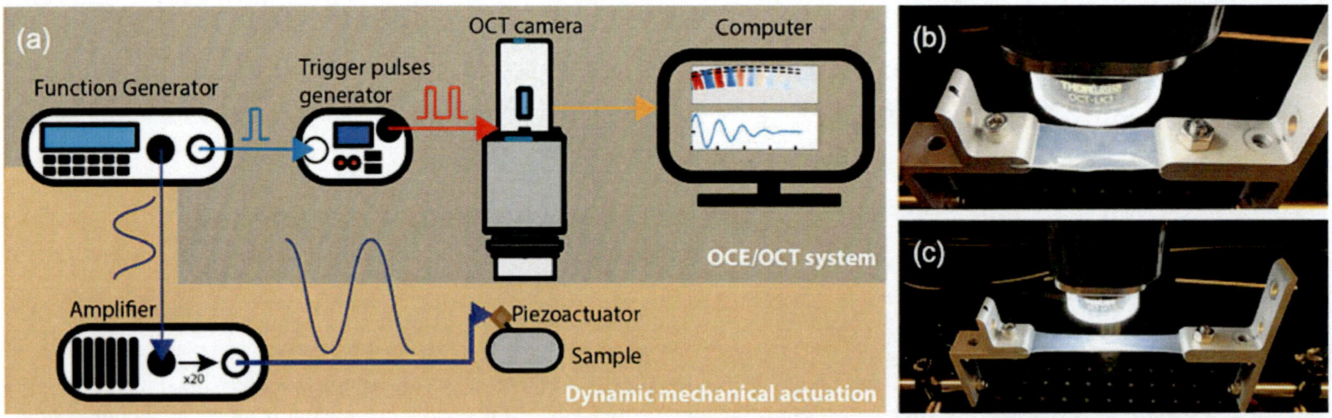

Fig. 1 Experimental setup with the optical coherence elastography/tomography (OCE/OCT) system and the dynamic mechanical actuation system. (**a**) The thin rubber membrane sample is mounted on the uniaxial stretch device under the OCT camera. The stretch ratio (λ_1) of the rubber sample is controlled by a precision translation stage between (**b**) $\lambda_1 = 1.0$ and (**c**) $\lambda_1 = 1.9$

stroma. To address this deficiency, first, we begin with a study of the influence of finite tensile deformation on the viscoelastic properties of a thin silicon rubber membrane subjected to uniaxial tension. Silicon rubber provides a surrogate model for the cornea tissue that eliminates the complexity associated with the anisotropic properties of corneal and the biaxial tensile state resulting from hydrostatic membrane loading. We can also study the biaxial tensile loading of the silicone membrane with an inflated membrane geometry. In this paper, we characterize the shear storage and loss moduli based on OCE measurements of the guided elastic wave velocity and attenuation in the rubber membrane. By applying uniaxial stretches between 1 and 1.9, we observe a corresponding modulation in the elastic wave speeds and wave amplitude decay, reflecting the membrane's stretch-dependent viscoelastic properties. We also explore a hyper-viscoelasticity framework for inverse analysis of the stretch-dependent elastic wave speed and attenuation to estimate the storage and loss moduli based on OCE measurements.

Background

In this section, we present a brief description of the OCT and OCE approaches and some experimental measurements in a silicone rubber membrane. Figure 1a shows a schematic illustration of the OCT/OCE experimental setup. The OCT microscope (Thorlabs GAN210C1, NJ, USA) is a low-coherence interferometer that transduces local variations in the morphology or displacement to optical intensity or phase signals. Elastic waves are generated by a paddle actuator and detected by the OCT interferometer. The actuator consists of a knife edge that is attached to the tip of a resonant piezoelectric (PZT) transducer (PA4CEW, Thorlabs) at one end. The other end of the knife edge is placed in contact with the sample surface. An amplified sinusoidal voltage from a function generator is used to drive the PZT transducer, leading to the local oscillatory tapping motion of the sample. Harmonic elastic waves generated in the sample by the oscillatory indentation of the sample propagate away from the actuation point, leading to local dynamic displacement in the sample. The OCT interferometer tracks the local out-of-plane displacement of the sample. Further details of the OCT/OCE approaches are available in our previous papers [18, 19]. In this work, we study the propagation of elastic waves in a thin silicone rubber membrane subjected to uniaxial tension. The polymer was made from EcoFlex 00-10 following the description in Royston et al. [20] The sample is mounted on a uniaxial stretching device and placed under the OCT camera, as shown in Fig. 1b, c.

Analysis

Figure 2a shows a two-dimensional OCT structural image of the sample for a stretch ratio of 1.0. The grayscale image corresponds to the backscattered intensity of the probe light, varying from black at the minimum level to white at the maximum. Two bright lines in the image occur at the top and bottom boundaries due to the large refractive index mismatch between the membrane and the surrounding air. The few dark vertical stripes within the sample are due to tiny air bubbles in the sample. The OCT image is uniform, suggesting that the sample is homogeneous. Figure 2b shows the corresponding OCE

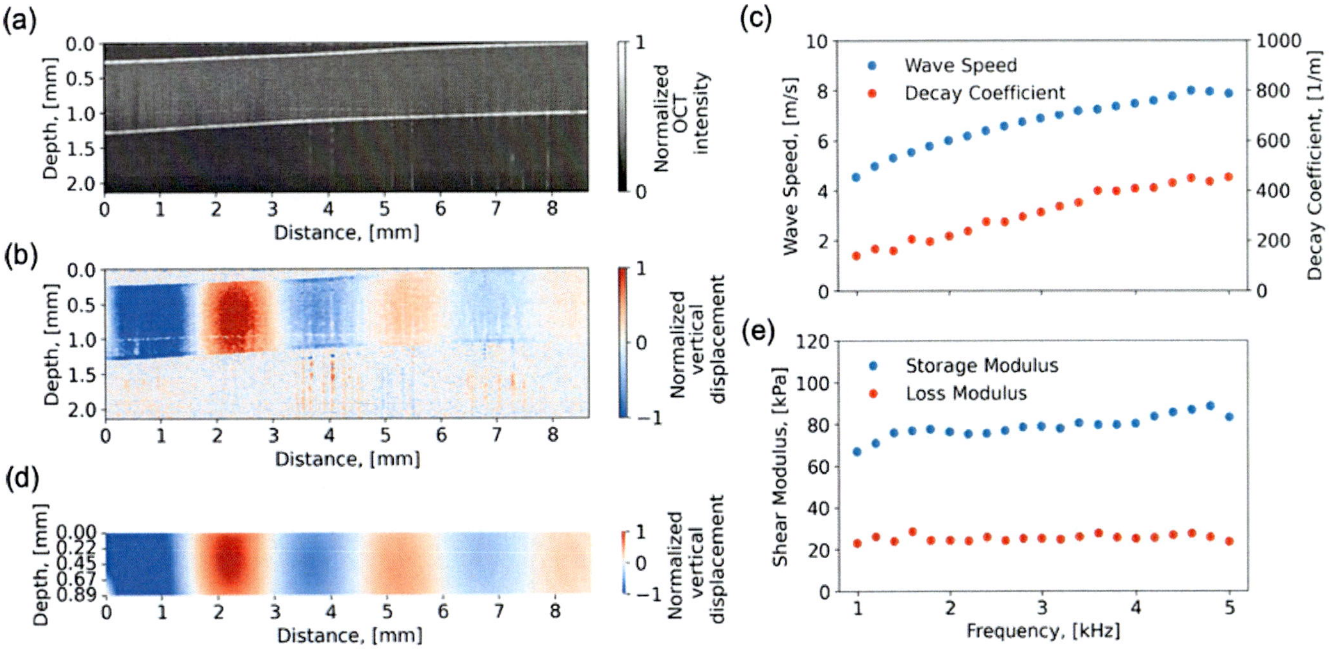

Fig. 2 Experimental results from the rubber membrane. (**a**) OCT image of the rubber membrane. (**b**) OCE image of the rubber membrane at 2 kHz. (**c**) The rubber membrane's wave speeds and decay coefficients from 1 to 5 kHz. (**d**) Numerically calculated vertical displacement field in the membrane for a harmonic excitation frequency of 2 kHz. (**e**) Estimated storage moduli and loss shear moduli for the rubber membrane

image. For the measurement, the PZT transducer was excited with a 2 kHz sinusoidal voltage. The periodic red-blue fringes in the image represent the spatial variation of the optical phase difference between the backscattered light and a reference light beam in the OCT interferometer. The phase difference is proportional to the local vertical displacement and refractive index [15]. The red-blue contrast in the image is due to the variation of the vertical displacement from positive (or upward moving) to negative (or downward moving) values. The local vertical displacement field is uniform through the depth of the sample. However, the displacement varies periodically in the lateral direction with a nearly constant wavelength. Additionally, the displacement decreases with distance from the excitation source. Similar behaviors are observed for excitation frequencies between 1 and 5 kHz, indicating that elastic waves within this range are attenuated with propagation distance due to a combination of material damping and diffraction.

To extract the wavelength and amplitude decay coefficient in the measured OCE data, first, we averaged multiple line scans from different depths in the OCE image to obtain a single displacement waveform. An exponentially decaying sinusoidal function, $I(t) = Ae^{-\alpha(f)x} \times \sin(2\pi x/\lambda(\mathrm{f}) - \theta)$, is fitted to the resulting displacement waveform, where $c(\mathrm{f}) = \lambda(\mathrm{f}) \times \mathrm{f}$ is the elastic wave speed, $\lambda(\mathrm{f})$ is the wavelength, and f is the elastic wave frequency. $\lambda(\mathrm{f})$ is equal to the spatial period of the fringe pattern in the OCE image, and θ is a phase constant. (f) is the amplitude decay coefficient $\alpha(f)$. To eliminate the effects of diffraction on α, we applied a diffraction correction factor to displacement waveform following the description in Ruiz and Nagy [21], since the width of the knife edge actuator (3 mm) is smaller than the longest propagation distance (i.e., ~ 9 mm) for the measured displacement field for the elastic waves in the membrane. The diffraction correction eliminates the displacement amplitude decay that stems from purely geometric effects like the finite size of the excitation source. Figure 2c shows the wave speeds and decay coefficients obtained from the displacement profiles for 1 to 5 kHz frequencies. The wave speed increases with frequency, going from 4.5 m/s at 1 kHz to 7.9 m/s at 5 kHz. The frequency-dependent wave speed (i.e., the phase velocity dispersion curve) is consistent with the theoretical dispersion curve for the membrane's zeroth-order A_0 antisymmetric Lamb wave mode in a thin plate [22]. Unlike Lamb waves in an elastic plate, the decay coefficient for the guided waves in the silicone membrane increases with the frequency from 140 to 450 m^{-1}, suggesting that the elastic wave dissipation increases with frequency.

We estimate the sample's frequency-dependent shear storage and loss moduli by fitting a visco-elastodynamic model for an incompressible thin silicone membrane to the measured elastic wave speed and amplitude decay coefficient. The thickness of the sample is estimated from the OCT image in Fig. 1a. The sample density is assumed to be 965 kg/m^3 [20]. Figure 2d shows the calculated displacement field in the membrane produced with a 2 kHz harmonic normal surface line load. The wavelength and amplitude decay ratio for displacement fields matches the diffraction-corrected experimental data. Figure 2e shows the best-fit frequency-dependent storage and loss shear moduli. Between 1 and 5 kHz, the shear storage modulus increases from 67 to 88 kPa. On the other hand, the shear loss modulus remains nearly constant and close to 25 kPa. The measured viscoelastic properties are being validated against low frequency shear rheometry measurements.

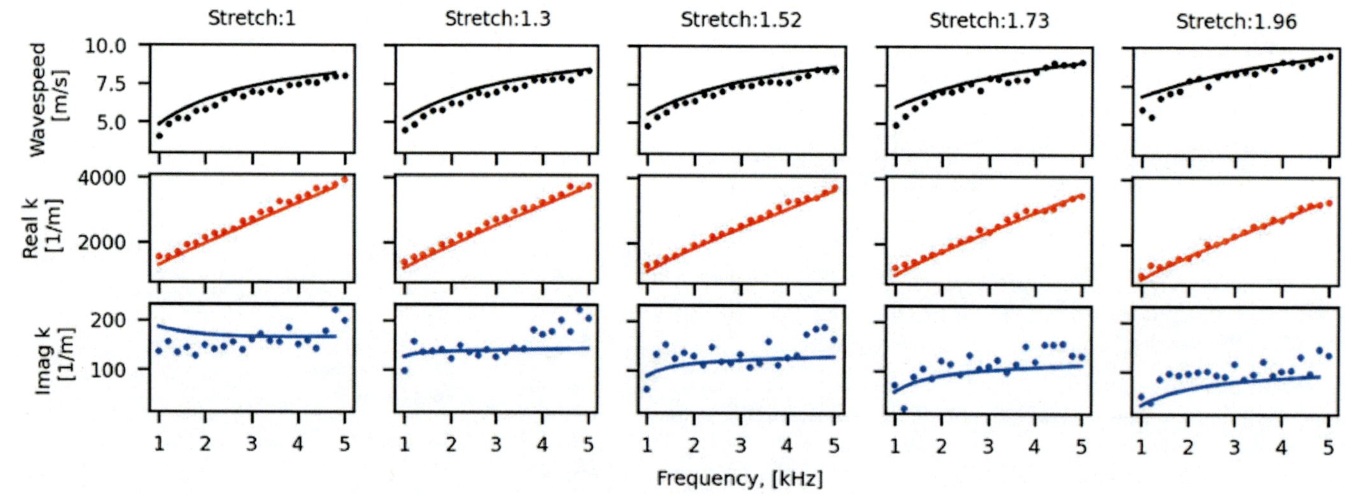

Fig. 3 OCE measurements (dots) and hyper-viscoelastic model (lines) of elastic wave speeds, real and imaginary components of the wavenumbers of guided waves in a silicone rubber membrane subjected to uniaxial stretch ratios from 1 to 1.96

Figure 3 shows the wave speeds and real and imaginary components of the wavenumbers for stretch ratios between 1 and 1.9. The real part of the wavenumber (k') is related to the wavelength (λ) by $k' = 2\pi/\lambda$, while the imaginary part of the wavenumber (k'') is the damping coefficient α. With an increasing stretch ratio, the wave speed increases, while the damping coefficient decreases. These results suggest that the material gets stiffer and less viscous with increasing uniaxial static deformation. To evaluate the membrane's stretch-dependent storage and loss modulus, we adopt the two-potential constitutive relation for rubber viscoelasticity developed by Kumar and Lopez-Paimes [23]. We employ the constitutive relation in the incremental equilibrium equations of motion [24] to calculate the frequency-dependent wave speed and amplitude decay for incremental guided wave motion that is superimposed on the finite deformation of the membrane. The preliminary numerical predictions from the model (solid line in Fig. 3) and the OCE measurements (dotted lines) are in good agreement, albeit that the goodness of the fits needs to be improved. The best-fit material properties, i.e., the stretch-dependent shear storage and loss moduli, will be reported in a later publication. Upon improving the quality of the data fitting, the modeling framework will be extended to investigate a hydrostatically loaded membrane in a state of biaxial tension and to simulate the effect of the hydrostatic pressure on the viscoelastic properties.

Conclusion

This work presents an experimental approach for viscoelastic characterization of soft membranes subjected to finite static deformations. Specifically, the experimental approach relies on measuring the frequency-dependent wave speed and amplitude decay coefficient for guided elastic waves in the membrane at different uniaxial stretch ratios. The preliminary results obtained for stretch ratios between 1 and 1.9 suggest that increasing the stretch ratio makes the membrane stiffer and less viscous. We incorporated a hyper-viscoelastic constitutive model for the stretched membrane into the elastodynamic equations to calculate the wave speed and amplitude decay coefficient of guided waves in the membrane. By matching these results to the measured data, we can estimate the stretch-dependent viscoelastic properties of the membrane. The proposed method can be used to study and characterize the viscoelastic behavior of soft membrane materials subjected to finite static deformations. In the future, we will explore the modeling and experimental approach to characterize the IOP dependence of the viscoelastic properties of corneal tissues.

References

1. Weinreb, R.N., Aung, T., Medeiros, F.A.: The pathophysiology and treatment of glaucoma: a review. JAMA. **311**(18), 1901–1911 (2014)
2. Robert, S.: A history of intraocular pressure and its measurement. Optom. Vis. Sci. **88**(1), E16–E28 (2011)
3. Mark, H.H.: Armand Imbert, Adolf Fick, and their tonometry law. Eye. **26**(1), 13–16 (2012)
4. Wessels, I.F., Oh, Y.: Tonometer utilization, accuracy, and calibration under field conditions. Arch. Ophthalmol. **108**, 1709–1712 (1990)

5. van der Jagt, L.H., Jansonius, N.M.: Three portable tonometers, the TGDc-01, the ICARE and the Tonopen XL, compared with each other and with Goldmann applanation tonometry*. Ophthalmic Physiol. Opt. **25**, 429–435 (2005)
6. Deol, M., Taylor, D.A., Radcliffe, N.M.: Corneal hysteresis and its relevance to glaucoma. Curr. Opin. Opthalmol. **26**(2), 96–102 (2015)
7. Susanna, C.N., Diniz-Filho, A., Daga, F.B., Susanna, B.N., Zhu, F., Ogata, N.G., Medeiros, F.A.: A prospective longitudinal study to investigate corneal hysteresis as a risk factor for predicting development of glaucoma. Am J. Ophthalmol. **187**, 148–152 (2018)
8. Liu, J., Roberts, C.J.: Influence of corneal biomechanical properties on intraocular pressure measurements: quantitative analysis. J. Cataract Refract. Surg. **31**, 146–155 (2005)
9. Guarnieri, F.A.: Corneal biomechanics. In: Corneal Biomechanics and Refractive Surgery, pp. 7–31. Springer, New York (2015)
10. Kohlhaas, M., Boehm, A.G., Spoerl, E., Pursten, A., Grein, H.J., Pillunat, L.E.: Effect of central corneal thickness, corneal curvature, and axial length on applanation tonometry. Arch. Opthalmol. **124**, 471–476 (2006)
11. Killer, H.E., Pircher, A.: Normal-tension glaucoma: review of current understanding and mechanisms of the pathogenesis. Eye (Lond.). **32**(5), 924–930 (2018)
12. Heijl, A., Leske, M.C., Bengtsson, B., Hyman, L., Bengtsson, B., Hussein, M., Early Manifest Glaucoma Trial Group: Reduction of intraocular pressure and glaucoma progression: results from the Early Manifest Glaucoma Trial. Arch. Ophthalmol. **120**(10), 1268–1279 (2002)
13. Anderson, D.R., Drance, S.M., Schulzer, M., Collaborative Normal-Tension Glaucoma Study Group: Factors that predict the benefit of lowering intraocular pressure in normal-tension glaucoma. Am J. Ophthalmol. **136**(5), 820–829 (2003)
14. Wolfs, R.C., Klaver, C.C., Vingerling, J.R., Grobbee, D.E., Hofman, A., de Jong, P.T.: Distribution of central corneal thickness and its association with intraocular pressure: the Rotterdam Study. Am J. Ophthalmol. **123**(6), 767–772 (1997)
15. Kirby, M.A., Pelivanov, I., Song, S., Ambrozinski, L., Yoon, S.J., Gao, L., Li, D., Shen, T.T., Wang, R.K., O'Donnell, M.: Optical coherence elastography in opthalmology. J. Biomed. Opt. **22**(12), 1–28 (2017)
16. Ambrozinski, L., Song, S., Yoon, S.J., Pelivanov, I., Li, D., Gao, L., Shen, T.T., Wang, R.K., O'Donnell, M.: Acoustic micro-tapping for non-contact 4D imaging of tissue elasticity. Sci. Rep. **6**, 1–11, 38967 (2016)
17. Pitre, Jr. J.J., Kirby, M.A., Li, D.S., Shen, T.T., Wang, R.K., O'Donnell, M., Pelivanov, I.: Nearly-incompressible transverse isotropy (NITI) of cornea elasticity: model and experiments with acoustic micro-tapping OCE. Sci. Rep. **10**, 1–14, 12983 (2020)
18. Liou, H.C., Sabba, F., Packman, A.I., Wells, G., Balogun, O.: Nondestructive characterization of soft materials and biofilms by measurement of guided elastic wave propagation using optical coherence elastography. Soft Matter. **15**, 575–585 (2019)
19. Liou, H.C., Sabba, F., Packman, A.I., Rosenthal, A., Wells, G., Balogun, O.: Towards mechanical characterization of granular biofilms by optical coherence elastography measurement of circumferential elastic waves. Soft Matter. **15**, 5562–5573 (2019)
20. Royston, T.J., Dai, Z., Chaunsali, R., Liu, Y., Peng, Y.: Estimating material viscoelastic properties based on surface wave measurements: a comparison of techniques and modeling assumptions. J. Acoust. Soc. Am. **1130**, 4126–4138 (2011)
21. Ruiz, M.A., Nagy, P.B.: Diffraction correction for precision surface acoustic wave velocity measurements. J. Acoust. Soc. Am. **112**, 835–842 (2002)
22. Simonetti, F.: Lamb wave propagation in elastic plates coated with viscoelastic materials. J. Acoust. Soc. Am. **115**, 2041–2053 (2004)
23. Kumar, A., Lopez-Pamies, O.: On the two-potential constitutive modeling of rubber viscoelastic materials. Comptes Rendus Mecanique. **344**, 102–112 (2016)
24. Ogden, R.W.: Incremental statics and dynamics of pre-stressed elastic materials. In: Waves in Nonlinear Pre-stressed Materials, pp. 1–26. Springer, Vienna (2007)

Open Source Contour Method Analysis for Assessing Residual Stress in Weldments

M. J. Roy, N. Stoyanov, and R. J. Moat

Abstract The contour method for assessing residual stress is a widely accepted technique and has been widely applied to validate and verify modelling predictions of welding processes since its inception 20 years ago. Briefly, it comprises the stress-free cutting of a body containing residual stress. Then, using the resulting averaged distortion measured from both cut faces, the calculation of the elastic loading to "force" the cut face "flat" is possible. However, a wider application as compared to other residual stress assessment techniques has been somewhat limited due to the absence of a common computational framework to process experimental data. This is at odds with the relatively accessible equipment requirements to generate this data: a wire electrodischarge machining (EDM) facility for cutting and a coordinate measurement machine or surface profilometer to capture the resulting deformation of the cut surfaces. In the present work, an open-source platform is presented that accommodates the full analysis work flow, from registration of surface profiles from each side of the cut, alignment and averaging of these profiles, numerical fitting of this surface, and finite element pre- and post-processing to recover a contour plot of the residual stress. Programmed in Python, this has been accomplished via a graphical user interface for initial preprocessing steps, followed by close integration of commercial finite element code (Abaqus) and open-source equivalents (Gmsh, CalculiX). Called pyCM (Python Contour Method), the overall analysis sequence is described employing a thick-section weld in pressure vessel steel (SA533), as well as showing that the results stemming from this tool employing either Abaqus or CalculiX are identical. Further advantages of pyCM will be discussed including diminished cost of accessing the technique and the ease of sharing of results and analysis strategies among practitioners. Factors, which necessitate these decisions, are inherent to the technique, and the present contribution lends itself to making the results obtained by the contour method more transparent.

Keywords Residual stress · FEA · Contour method · Welding · Software

Introduction

The contour method (CM) to experimentally assess residual stress in steel weldments for nearly two decades since being invented by Prime [1, 2]. The results obtained can be employed for fitness-for-service assessments, as well as validating computational process models. Although being a destructive technique, in that the component to be assessed is destroyed along with the residual stress to be quantified, it is relatively inexpensive to implement from an experimental standpoint. A detailed review of the technique has been published by Prime and DeWald [3], which has been closely followed by a best-practice guide covering both experimental and analytical approaches by Hosseinzadeh et al. [4]; we direct the reader to these for a background on the technique. Focusing on processing of the data that results from a contour method analysis,

M. J. Roy (✉) · N. Stoyanov
Henry Royce Institute, Department of Mechanical, Aerospace and Civil Engineering, University of Manchester, Manchester, UK
e-mail: matthew.roy@manchester.ac.uk

R. J. Moat
School of Engineering & Innovation, The Open University, Milton Keynes, UK

C. Franck et al. (eds.), *Challenges in Mechanics of Biological Systems and Materials, Thermomechanics and Infrared Imaging, Time Dependent Materials and Residual Stress, Volume 2*, Conference Proceedings of the Society for Experimental Mechanics Series,
https://doi.org/10.1007/978-3-031-50470-9_4

early approaches have been made publicly available, which employ MATLAB[1] and Abaqus[2] [5]. However, these approaches require programming alterations between datasets or instruments and do not incorporate the developments in analysis and best practices that have been developed in the past decade. Commercial or closed source software has been historically employed to conduct an analysis.

In the latter steps of the method, analytical approaches have been employed by Kartal et al. [6, 7] to solve for stresses directly from processed metrology data. However, finite element analysis (FEA) is nominally employed as it can be adapted to different geometries. Furthermore, the results of an FEA approach can be more readily transferred or compared to companion, component-level finite element analyses. For example, the European Network on Neutron Techniques Standardization for Structural Integrity (NeT) have employed the FEA-derived CM results to validate continuum weld models, examining the effects of constraint and how residual stresses develop [8]. However, the current status quo of the contour method has demonstrated that different practitioners can obtain different results.

Building on the advancements in open-source numerical tools, an open-source tool for conducting a CM analysis has been developed. Called pyCM (python Contour Method) [9], it encompasses the full workflow of analysis steps from accepting raw metrology data through to a FEA pre- and post-processor. The following will briefly introduce the software and demonstrate its use on a 80-mm-thick, narrow-gap TIG-welded S533 component which was produced as part of the New Nuclear Manufacturing (NNUMAN) program [10].

Overall Architecture

The overall pyCM analysis approach is presented in Fig. 1 relative to the analysis steps inherent to the CM. Each step in the analysis can be called independently as a stand-alone GUI or coupled together into a comprehensive tabbed interface. Furthermore, all steps/records are held in a single file, doing away with the need to separately archive FEA input and output files.

Historically, the flexibility required for CM surface analysis has lent itself well to an interpreted language, and MATLAB has been used extensively. However, there are shortcomings to employing MATLAB:

- The proprietary nature of the built-in libraries which impedes access.
- Although "compiled" code can be run on computers without being installed, this is reliant on the target machine having an identical runtime installation, which is proprietary.
- Diminished text file parsing makes working with FE codes difficult; this is particularly true for very large input files or mesh data.

Each of these factors restricts the overall portability and maintenance of the existing codes and presents barriers to the residual stress community to develop and validate analysis code.

While the proprietary aspect of MATLAB has been largely addressed by GNU Octave [11], a free and open-source superset of MATLAB, Octave is not optimized in terms of memory allocation and interpreter, focusing instead on compatibility. For this reason, many software developers use Python entirely for numerical analysis. Python is a general purpose, interpreted language, with many packages available to perform the same functions as MATLAB in both research and teaching [12]. By employing the Numerical Python [13] and SciPy [14] libraries, the same functionality is available as in MATLAB. Furthermore, there is a maintained library of Python bindings for VTK, the Visualization Toolkit [15]. This is a robust tool for visualizing large datasets. For example, ParaView, an open-source FEA post-processor, makes extensive use of VTK [16]. Finally, there are also Python bindings for the QT graphical user interface (GUI) library. All of these relatively recent developments have presented the opportunity to develop robust numerical calculation and visualization capabilities that can be employed with little restriction. Beyond the numerical and visualization requirements of CM analyses is the requirement for a linear elastic FEA. Typically, this is carried out by using a general purpose commercial code, Abaqus. CalculiX is a free and open-source package offering a linear elastic solver with input and outputs syntactically similar to Abaqus [17].

The overall architecture of pyCM, as well as the libraries and external routines which it employs, has been selected to promote collaboration. Migration from existing routines and procedures currently in use has been promoted in its realization. All subsequent figures are screen captures directly from the pyCM interface with minor annotations applied.

[1] Registered trademark of The MathWorks, Inc.

[2] Registered trademark of Dassault Systèmes.

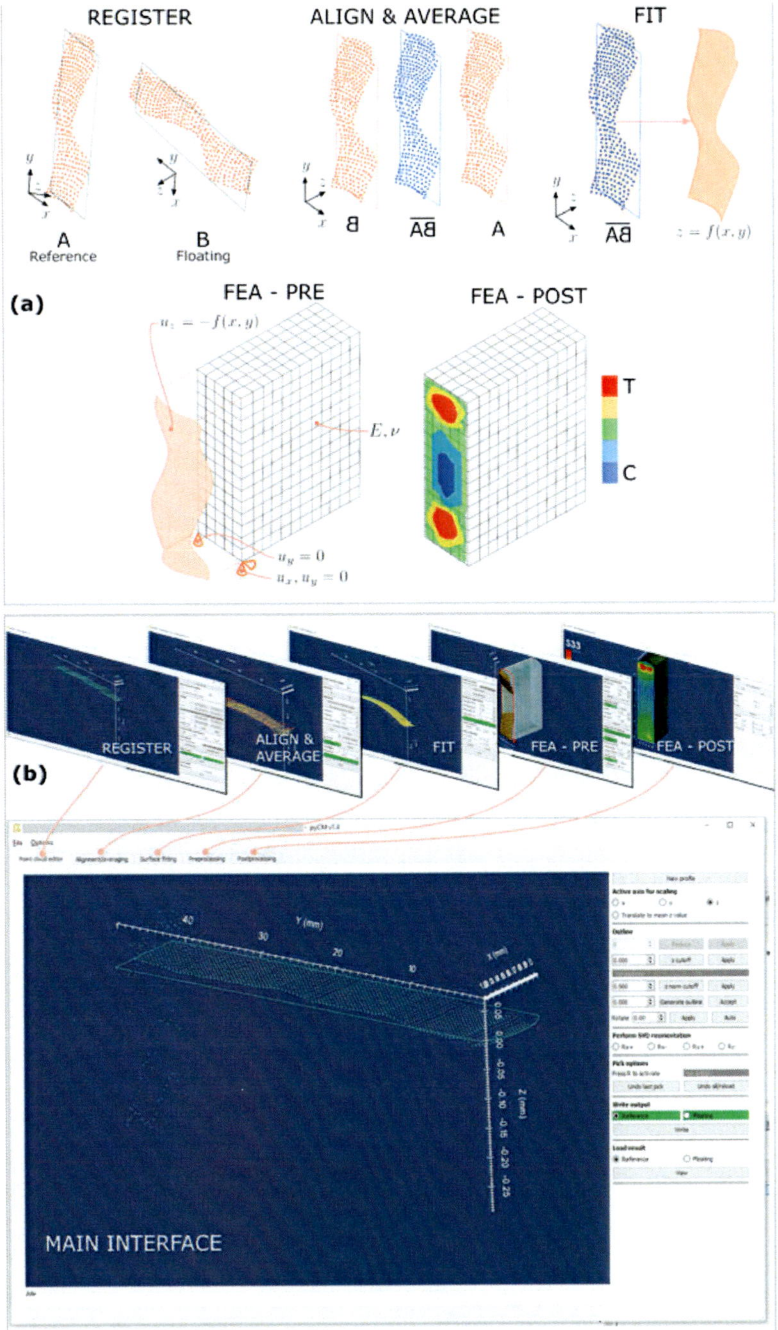

Fig. 1 The CM approach (a) and realization with pyCM (b)

Case Study

In order to demonstrate the analysis technique, a narrow-gap gas tungsten arc welding procedure was applied to two SA533 Gr.B Type II low alloy steel plates in the configuration shown in Fig. 2. A complete description of the welding protocols used to fabricate this component has been published by Rathod et al. [18] which details the metallurgical effects of welding on mechanical properties of the weld. However, a second motivation was to ensure that SA508 Grade 3 Class 1 welding trials could be carried out without overloading the designed restraint features. To that end, a contour cut was performed on the as-welded plates, and the resulting surfaces were measured prior to metallurgical and mechanical coupons being extracted.

The measurement methodology with a coordinate measurement machine (CMM) produced outlines/perimeters of components with a pitch of 0.2 mm and approximately 11,000 points per side, with an adaptive measurement technique that provided twice the point density at approximately the weld centerline than elsewhere. This data was then read into

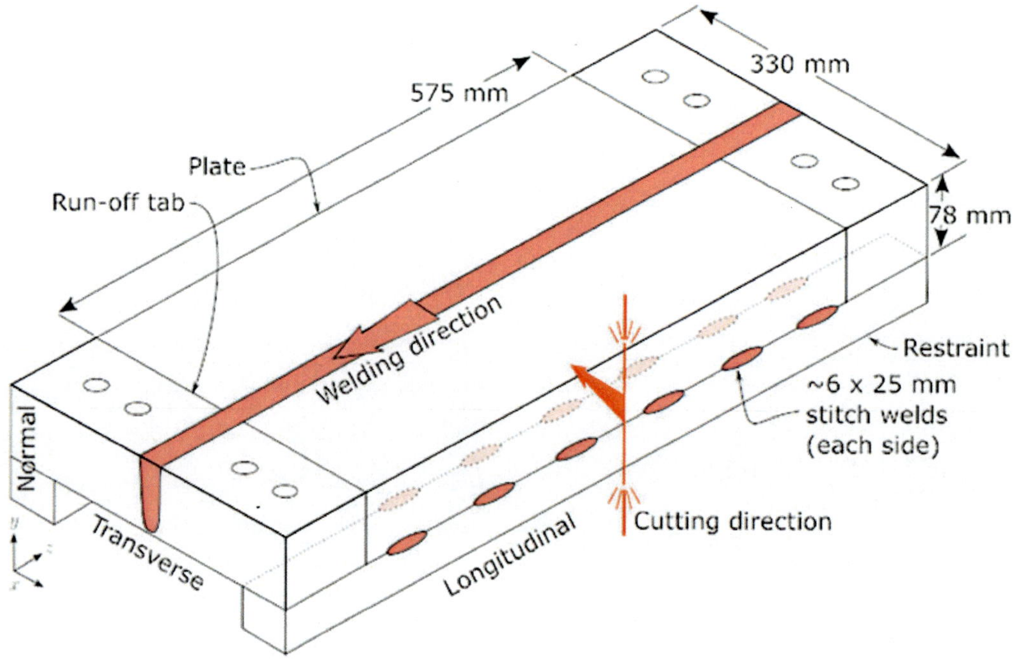

Fig. 2 Configuration of SA533 welding trial and CM cut location. Restraint bars were retained through the CM cut

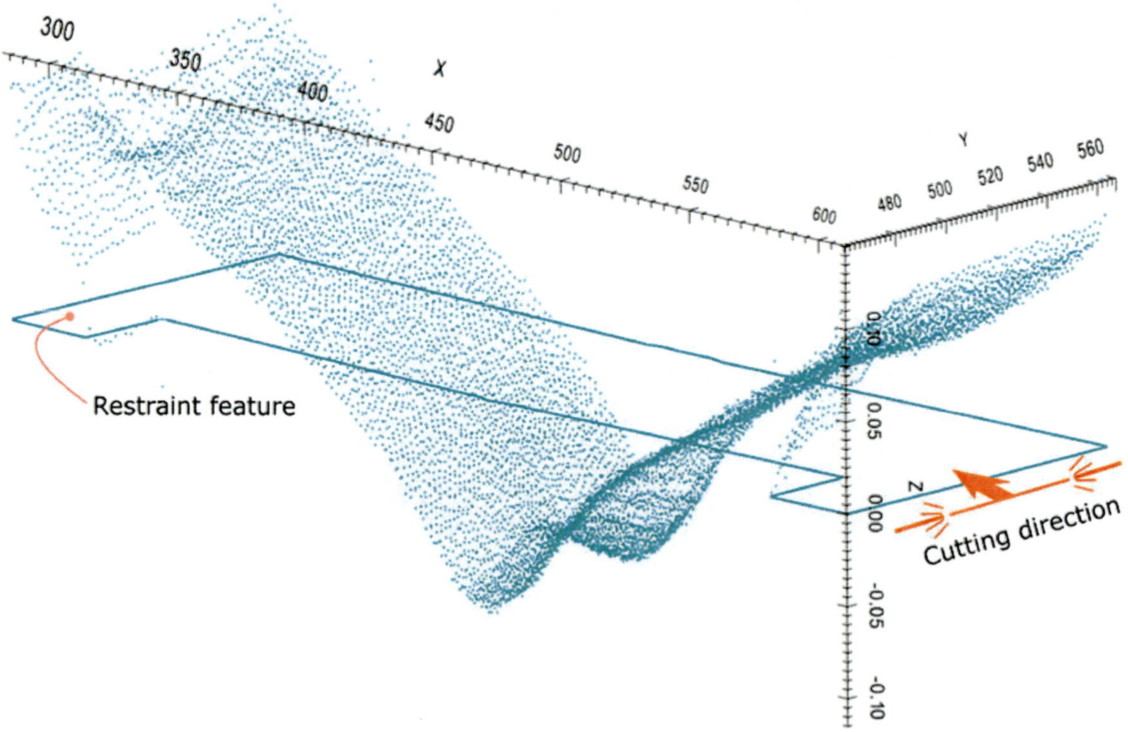

Fig. 3 Reference point cloud and corresponding outline

pyCM v1.1.8, with one side set as a reference and the other "floating." No sensitivity to the final result was noted when these were reversed. See Fig. 3 for the rendered reference dataset. This was accomplished with the "register" interface (Fig. 1).

The floating dataset was then mirrored, and the outline/perimeter was then used to align it with the outline of the reference dataset using the "align & average" interface (Fig. 1). This was done using an iterative closest-point algorithm, with minor manual adjustments. Once aligned, the reference and floating point clouds were averaged using linear interpolation to generate an averaged point cloud, arranged equidistant from each other in the *xy* plane. Points on this grid falling outside the reference outline were automatically discounted. The result is shown in Fig. 4.

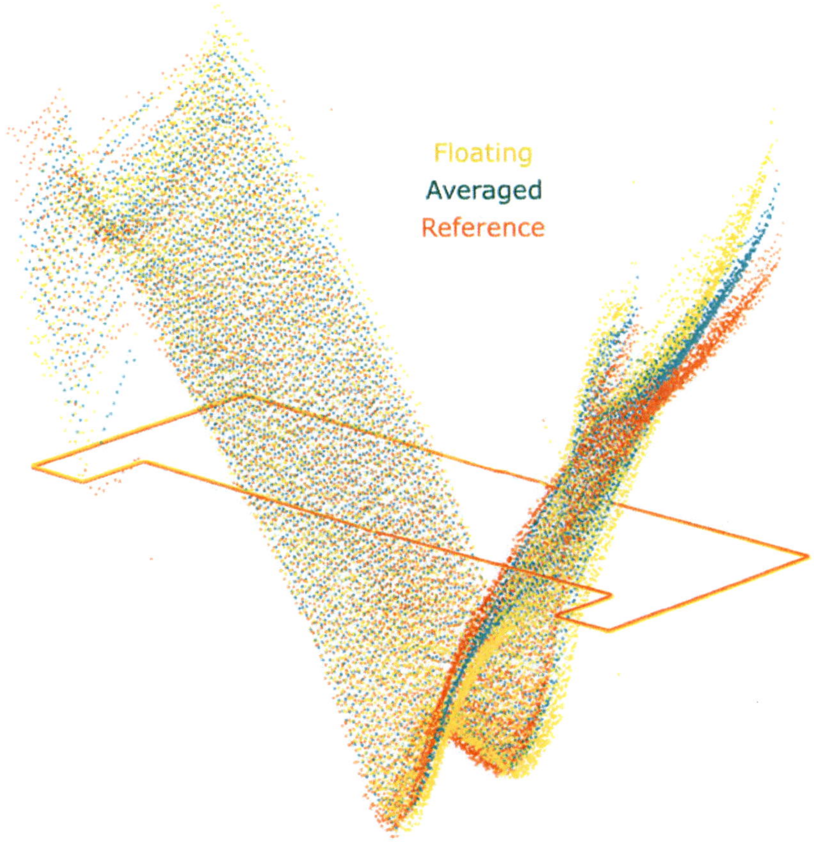

Fig. 4 Aligned and averaged point cloud

In the next step, the averaged point cloud was fitted with a cubic bivariate spline, with an equidistant knot spacing at 8 mm along the x and y directions ("fit" in Fig. 1). The interface employs the FITPACK algorithm available through SciPy, whereby the user has the option to set the spline order and uniform knot spacing independently in the x and y directions. It was found that the above knot spacing produced the best fit without "overfitting" in both directions (Fig. 5a), as assessed qualitatively. In order to assess this, the windowing function was employed in the spline fitting GUI to show bands of points and the spline's fit through this region (Fig. 5b).

With a spline now fitted through the averaged point cloud, the FEA preprocessing step ("FEA-pre" – Fig. 1) was then available. This started first with seeding of the reference outline where 300 seeds were specified, resulting in approximately 3 mm line segments. This informed a Gmsh script generated by pyCM, along with an extrusion length of 288 mm and 10 partitions to be geometrically placed along this length to produce the mesh shown in Fig. 6. The nodes on the front face of the mesh were then identified by pyCM, and displacement and rigid boundary conditions were imposed. The latter were specified from candidates presented to the user. Upon specifying a modulus of 200 GPa and Poisson's ratio of 0.3, a final FEA input file was generated by pyCM suitable to be submitted to the CalculiX or Abaqus solver. While subsequent results were produced by CalculiX, there was no difference in final results between employing Gmsh and an Abaqus solver or Gmsh and the CalculiX solver. The same was found to be true when employing Abaqus meshing in place of Gmsh.

The resulting FEA output files were parsed and the results displayed in the final pyCM step ("FEA-post" – Fig. 1). Near-yield longitudinal stresses were found within the top half of the weld region, balanced by high compressive residual stresses in the regions proximate to the restraint features as shown on the top pane of Fig. 7. Minor wire entry/exit artifacts were found on the inner surfaces of the restraint features, which was expected as the cut cross-section changed appreciably in these locations. Probe lines across the regions of weld with the highest and lowest residual stress were extracted (bottom pane of Fig. 7) with the pyCM utility, and this shows that while the stress at approximately a third the way through the weld is low, it still is tensile throughout.

The arrangement and distribution of longitudinal stress are the result of the multiple-pass welding technique employed, whereby subsequent passes served to temper out the residual stresses developed by previous passes. Beyond this observation, the main observation is that the stresses developed in the restraint features remained below the yield point. While the

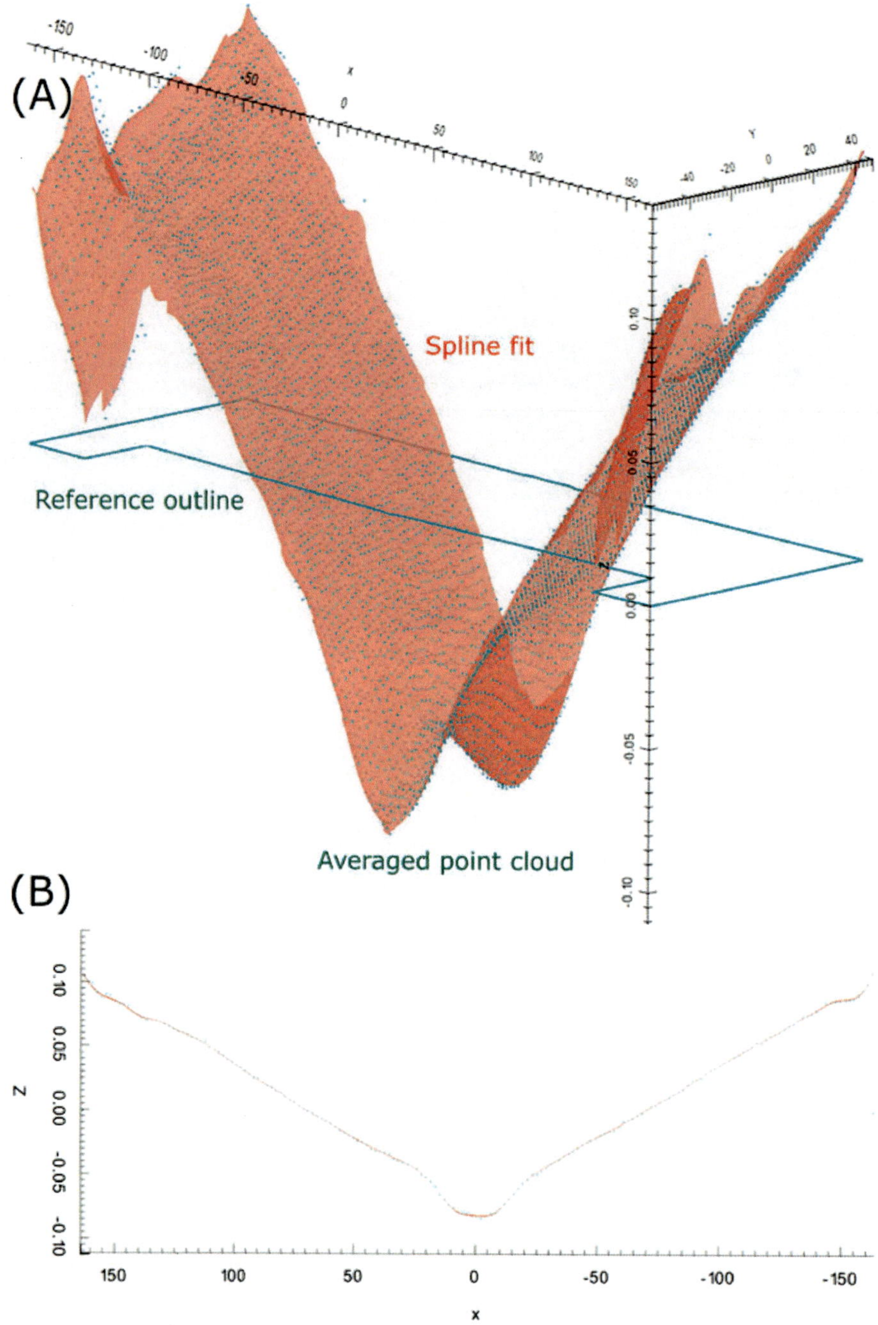

Fig. 5 Averaged point cloud fit with a bivariate cubic spline (**a**), resulting segment of points falling between $-1 > y > 1$ mm and the resulting spline (**b**)

preceding analysis would benefit from further validation via diffraction or another complimentary approach, it is surmised that it sufficiently demonstrates the ability of pyCM to produce plausible and repeatable results.

Discussion

Employing a series of open-source libraries, the pyCM framework has demonstrated capable of performing analyses according to published best practice. However, there have been several advances in the practical implementation of the overall technique that remain to be included in pyCM.

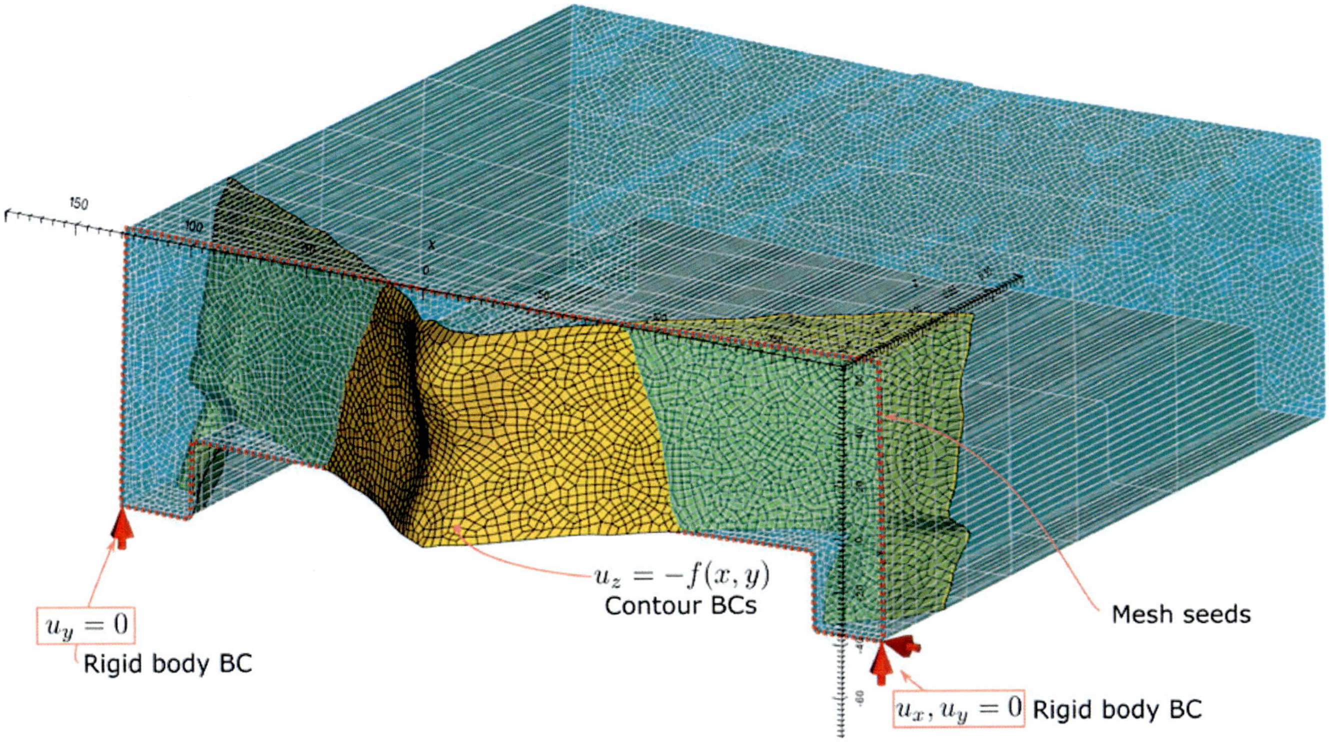

Fig. 6 Preprocessed FEA displaying mesh details, rigid body, and displacement boundary conditions imposed

First, cutting strategies incorporating sacrificial material bonded to the sample prior to sectioning in order to avoid cutting artifacts as the cut length varies sharply for irregular components (such as the component presented in the preceding section) need to be accounted for. With this approach, no change to the pyCM workflow is required. However, a complementary approach would be to perform multiple cuts, such that two contour method analyses are carried out: one on the body (i.e., in the residually stressed state) and a subsequent one parallel close to the first where the residual stress will have largely relaxed. The "stress" obtained from this second "stress-free" cut could then be subtracted from the stresses obtained for the first cut to obtain the underlying residual stress by superposition. This approach parallels that employed for diffraction techniques, whereby a sample free from macroscopic stress ("stress-free") is necessary to provide a datum. Currently, pyCM supports this approach but requires separate analyses across multiple data files, and this could be adapted relatively easily in the future.

On the processing side, particularly for surface fitting, there are other techniques available to generate a numerically continuous surface. While a bi-variate spline has shown to be quite flexible, there are other approaches such as local polynomial regression and Fourier series fitting [19]. The flexibility of such approaches to non-rectangular domains remains a challenge; however there are advantages in terms of processing data obtained from interrupted cutting, such as that generated when employing a self-restraint technique to avoid cutting-induced plasticity [20, 21]. The fitting component of pyCM could be changed to support complementary surface fitting. For example, large wavelength stresses could be fitted with a Fourier series or non-uniform, bivariate splines, and shorter wavelengths with bivariate, regular splines.

Third, there have been advancements on cutting of components which are not symmetric about the cutting plane [22]. In these circumstances, an erroneous result may be obtained as the stiffness is different on each side of the cut. The approach taken is to apply the contour boundary conditions on a geometry which as closely as possible represents each side and then average the resultant stress. Examples of where this might occur are in components which are not symmetric about the cutting plane over a short distance – the end of a weld, for example. In order to best accommodate these approaches, more flexible meshing could be introduced into pyCM and the facility to import meshes generated elsewhere, along with the facility to align contour boundary conditions to these external meshes. Fortunately, expanding the FEA pre- and post-processing facility within pyCM incorporating this approach will permit multiple cuts of the same component to assess stresses acting on other planes.

Finally, due to the numerous parameters inherent to contour method analysis, the best methodology for assessing the accuracy of an application remains measurement with a different approach, such as diffraction. While routes for error assessment due to numerical fitting of a contour surface have been proposed [19], a facility for assessing mesh sensitivity

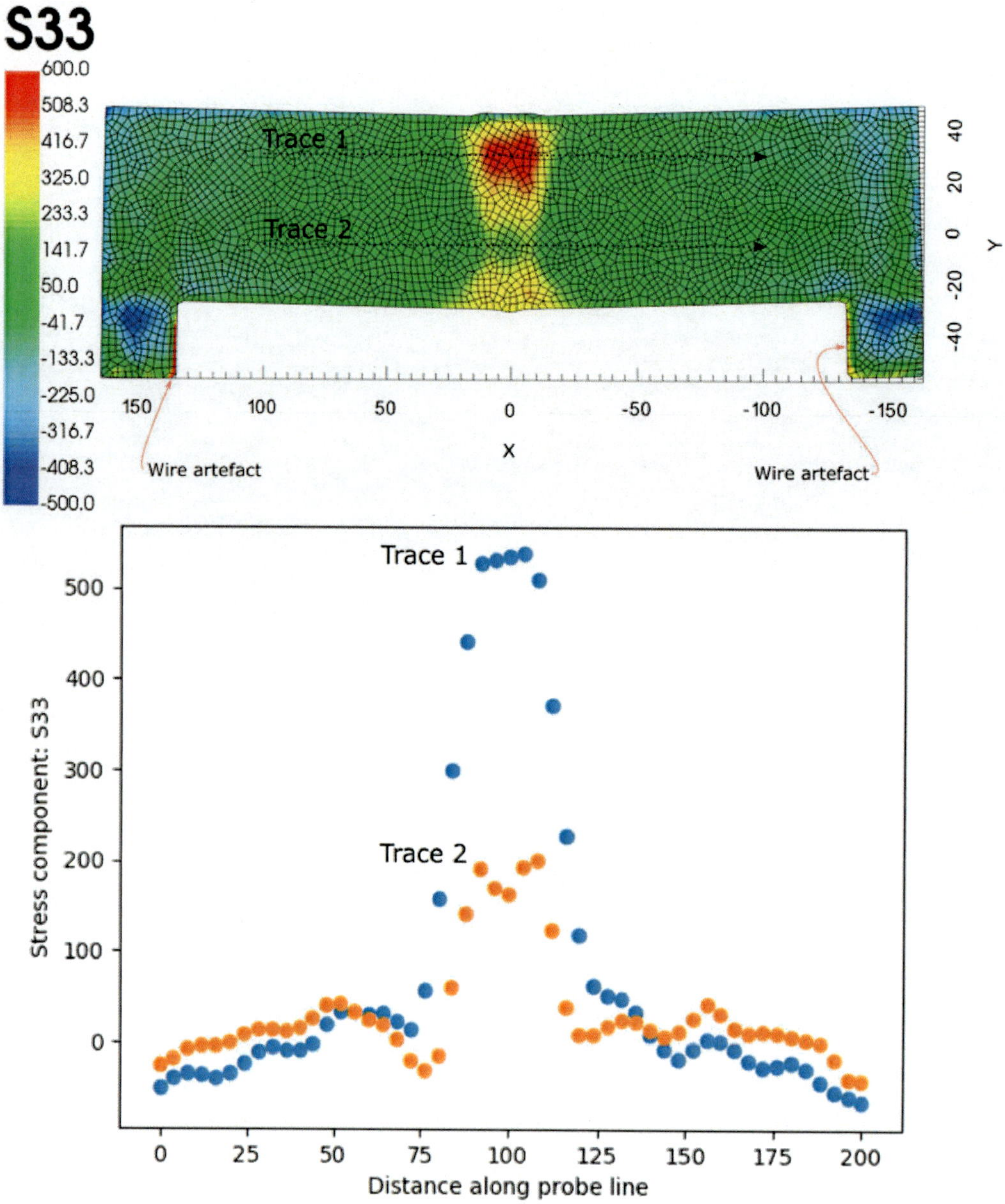

Fig. 7 Post-processed results demonstrating the "line probe" functionality to extract longitudinal residual stresses in MPa

and multiple fitting function types remains a largely manual process. However, pyCM can be adapted to address this with the ability to support multiple analyses within a single data file.

Summary

An open-source tool that encompasses all of the steps required to perform a contour method analysis has been developed, and its application has been demonstrated on a thick section steel weldment containing near-yield residual stresses. It is confirmed that the same results are obtained for the same mesh regardless a free and open-source finite element package (CalculiX) is employed or a commercial one, Abaqus.

Moving forward, the intention is to continue the development of pyCM to permit advancements in the application of the contour method. These include new methods for dealing with sacrificial materials and/or stress-free cuts, interrupted cutting, and more complicated specimen geometries. For improving interpretation of the contour method results, a tool for automating sensitivity analyses on surface fitting parameters and mesh density has also been identified as a priority.

Acknowledgments M.J. Roy and N. Stoyanov would like to acknowledge the financial support from the EPSRC (EP/L01680X/1) through the Materials for Demanding Environments Centre for Doctoral Training. M.J. Roy would also like to thank D. Rathod for making the contour cut data available. We are thankful to the late Dr. Mike Fox for his initial efforts in the development of the subject software.

References

1. Prime, M.B.: Cross-sectional mapping of residual stresses by measuring the surface contour after a cut. J. Eng. Mater. Technol. **123**(2), 162–168 (2001)
2. Prime, M.B., Sebring, R.J., Edwards, J.M., Hughes, D.J., Webster, P.J.: Laser surface-contouring and spline data-smoothing for residual stress measurement. Exp. Mech. **44**(2), 176–184 (2004). https://doi.org/10.1007/BF02428177
3. Prime, M.B., DeWald, A.T.: The contour method. In: Practical Residual Stress Measurement Methods, pp. 109–138. Wiley-Blackwell (2013). https://doi.org/10.1002/9781118402832.ch5
4. Hosseinzadeh, F., Kowal, J., Bouchard, P.J.: Towards good practice guidelines for the contour method of residual stress measurement. J. Eng. [Online] (2014). Available: http://digital-library.theiet.org/content/journals/10.1049/joe.2014.0134
5. Johnson, G.: Residual Stress Measurements Using the Contour Method. University of Manchester, Manchester (2008)
6. Kartal, M.E.: Analytical solutions for determining residual stresses in two-dimensional domains using the contour method. Proc. R. Soc. A. **469**(2159), 20130367 (2013)
7. Kartal, M.E., Kang, Y.-H., Korsunsky, A.M., Cocks, A.C.F., Bouchard, J.P.: The influence of welding procedure and plate geometry on residual stresses in thick components. Int. J. Solids Struct. **80**, 420–429 (2016). https://doi.org/10.1016/J.IJSOLSTR.2015.10.001
8. Smith, M.C., Smith, A.C., Wimpory, R., Ohms, C.: A review of the NeT Task Group 1 residual stress measurement and analysis round robin on a single weld bead-on-plate specimen. Int. J. Press. Vessel. Pip. **120–121**, 93–140 (2014). https://doi.org/10.1016/J.IJPVP.2014.05.002
9. Roy, M.J., Stoyanov, N., Moat, R.J., Withers, P.J.: pyCM: an open-source computational framework for residual stress analysis employing the Contour Method. SoftwareX. **11**, 100458 (2020). https://doi.org/10.1016/j.softx.2020.100458
10. Jeyaganesh, B., et al.: Overview of welding research under the New Nuclear Manufacturing (NNUMAN) programme. In: American Society of Mechanical Engineers, Pressure Vessels and Piping Division (Publication) PVP, vol. 6B, (2014). https://doi.org/10.1115/PVP2014-29015
11. Eaton, J.W.: GNU Octave Manual. Network Theory Limited, New York (2002)
12. Fangohr, H.: A comparison of C, MATLAB, and python as teaching languages in engineering. In: Computational Science – ICCS 2004, pp. 1210–1217. Springer, Berlin, Heidelberg (2004)
13. Oliphant, T.E., et al.: A guide to NumPy, vol. 1. Trelgol Publishing, USA (2006)
14. Oliphant, T.E.: Python for scientific computing. Comput. Sci. Eng. **9**(3), 10–20 (2007)
15. Schroeder, W., Martin, K.M., Lorensen, W.E.: The Visualization Toolkit an Object-Oriented Approach to 3D Graphics. Prentice-Hall, Inc., Upper Saddle River (1998)
16. Ahrens, J., Geveci, B., Law, C., Hansen, C., Johnson, C.: Paraview: an end-user tool for large-data visualization. Vis. Handb. **717**, 50031–50038 (2005)
17. Dhondt, G.: The Finite Element Method for Three-Dimensional Thermomechanical Applications. John Wiley & Sons, Chichester, West Sussex (2004)
18. Rathod, D.W., Sun, Y., Obasi, G., Roy, M.J.: Effect of multiple passes on Lüders/yield plateaus, microstructure and tensile behaviour of narrow-gap thick-section weld plates. J. Mater. Sci. **54**(19) (2019). https://doi.org/10.1007/s10853-019-03822-z
19. Olson, M.D., DeWald, A.T., Prime, M.B., Hill, M.R.: Estimation of uncertainty for contour method residual stress measurements. Exp. Mech. **55**(3), 577–585 (2015). https://doi.org/10.1007/s11340-014-9971-2
20. Hosseinzadeh, F., Traore, Y., Bouchard, P.J., Muránsky, O.: Mitigating cutting-induced plasticity in the contour method, part 1: experimental. Int. J. Solids Struct. **94–95**, 247–253 (2016). https://doi.org/10.1016/J.IJSOLSTR.2015.12.034
21. Muránsky, O., Hamelin, C.J., Hosseinzadeh, F., Prime, M.B.: Mitigating cutting-induced plasticity in the contour method. Part 2: numerical analysis. Int. J. Solids Struct. **94–95**, 254–262 (2016). https://doi.org/10.1016/J.IJSOLSTR.2015.12.033
22. Achouri, A., Hosseinzadeh, F., Bouchard, P.J., Paddea, S., Muransky, O.: The incremental contour method using asymmetric stiffness cuts. Mater. Des. **197**, 109268 (2021). https://doi.org/10.1016/j.matdes.2020.109268

Variation of Initial Bulk Residual Stresses in Aluminum Alloy 7050-T7451 and Its Effect on Distortion of Thin-Walled Structural Parts

Daniel Weber, Benjamin Kirsch, Nicholas A. Bachus, Christopher R. D'Elia, Barbara S. Linke, Michael R. Hill, and Jan C. Aurich

Abstract Milled thin-walled monolithic aluminum structural parts are widely used in the aerospace industry due to their appropriate properties such as a high overall strength-to-weight ratio. The semifinished products made of aluminum alloy 7050 undergo a heat treatment to gain this increased strength and hardness: Typically, three steps including solution heat treatment, quenching, and age hardening are carried out. This also leads to high initial bulk residual stresses (IBRS) within the part in the range of ±200 MPa. In rolled aluminum plate, plastic stretch (T7451 heat treatment designation) provides very effective relief of residual stress, decreasing IBRS to a level of ±20 MPa. In large parts with low bending stiffness, even that low level of IBRS can cause appreciable distortion. Besides the IBRS, the machining-induced residual stresses (MIRS) contribute to the distortion. This study investigates how IBRS in different stress relieved 7050-T7451 semifinished products vary and how this variation affects the distortion of milled thin-walled monolithic structural parts. A linear elastic finite element distortion prediction model, which considers the IBRS as well as the MIRS as input, was used to analyze the effect of varying IBRS on the distortion for different part sizes and geometries. The model was validated by machining of those parts and measuring their distortion. IBRS were measured via slitting technique and MIRS via incremental hole-drilling.

Keywords Residual stress · Distortion · Milling · Aluminum 7050-T7451 · Finite element modelling

Introduction

Aluminum alloys appear as structural materials in various industries such as aerospace, automotive, electrical, and chemical [1]. For example, the aerospace industry demands a high mass fraction of aluminum alloys in airplanes. Especially thin-walled, monolithic structural components are used due to their beneficial properties, such as a high overall strength-to-weight ratio and good fatigue life [2]. Typically, those structural components are machined by the manufacturing process milling, removing up to 95% of the total volume of the semifinished product [3]. Part distortion is a common problem due to the required small wall thicknesses of those weight-optimized structural components [4]. Residual stresses (RS) in the form of machining-induced residual stresses (MIRS) and initial bulk residual stresses (IBRS) are known to be the main cause of those distortions [4]. The MIRS are induced by the machining process due to plastic deformations in a near surface layer of the part. Prior research has investigated the effects of cutting conditions and tool properties on MIRS [5–8]. The IBRS are present in the blank material because of upstream processes like rolling, casting, and especially the heat treatment, which is characterized for wrought and cast aluminum alloys by a three-step process to increase the material strength and hardness: solution heat treatment, quenching, and age hardening [9]. Especially the high thermal gradients during quenching lead to high IBRS in the range of −150 to +100 MPa within the material. Therefore, a RS relief through controlled stretching or compression is necessary. Although the RS can be reduced to a level of ±20 MPa or lower, distortion remains as a problem. Past investigations showed that the position of the finished part in the semifinished product has an influence on the part distortion due to the redistribution of the IBRS [10–14]. However, it was not investigated yet how IBRS in different batches of A7050-T7451 vary and if that affects the distortion of thin-walled structural parts.

D. Weber (✉) · B. Kirsch · J. C. Aurich
Institute for Manufacturing Technology and Production Systems, RPTU in Kaiserslautern, Kaiserslautern, Germany
e-mail: daniel.weber@rptu.de

N. A. Bachus · C. R. D'Elia · B. S. Linke · M. R. Hill
Department of Mechanical and Aerospace Engineering, University of California, Davis, CA, USA

C. Franck et al. (eds.), *Challenges in Mechanics of Biological Systems and Materials, Thermomechanics and Infrared Imaging, Time Dependent Materials and Residual Stress, Volume 2*, Conference Proceedings of the Society for Experimental Mechanics Series,
https://doi.org/10.1007/978-3-031-50470-9_5

Numerical simulation models, typically based on the finite element method (FEM), are the preferred choice for predicting the part distortion due to known RS, because of the reduced effort of the modelling process especially for complex part geometries in comparison to analytical methods. The RS used as input for the FEM distortion model are either measured [14–16] or predicted analytically or numerically [2]. The FEM distortion models are mostly based on linear material elasticity. The modelling of the material removal process, e.g., realized by element deletion techniques [15, 16] or the application of the RS to the final machined part geometry [2, 14], was presented in the literature. Either way, forces and moments act as a consequence of the RS imbalance, which leads to the distortion of the part due to the re-equilibrium of the RS as soon as the component is released from its constraints (clamping).

In our past research, we investigated the effect of varying MIRS on distortion [17]. Furthermore, we developed a FEM model to predict the part distortion due to both MIRS and IBRS [18–20]. This study investigates how the IBRS in different 7050-T7451 semifinished products vary. With the help of the FEM distortion model, we assess how this variation affects distortion of structural parts.

Methods

Prior to investigating the effect of IBRS on distortion, the IBRS in three different configurations of material A7050-T7451 were measured by the slitting method. Then, machining of different part geometries and measuring their distortion was done to validate predictions from the FEM distortion model.

Initial Bulk Residual Stress Measurements

To investigate the effect of IBRS on distortion, IBRS in three different A7050-T7451 materials were determined:

- A: Vendor I (USA; 102-mm-thick plate): 206 × 28 × 102 mm^3 (x-direction is longitudinal (L), rolling direction; y-direction is long-transverse (LT) direction; z-direction is short transverse (ST) direction)
- B: Vendor II (Ger; 30-mm-thick plate) batch i: 206 × 102 × 30 mm^3 (x- is L direction; y- is LT direction; z- is ST direction)
- C: Vendor II (Ger; 30-mm-thick plate) batch ii: 206 × 102 × 30 mm^3 (x- is L direction; y- is ST direction; z- is LT direction)

For two material batches, B and C, small beam specimens (see Fig. 1a), measuring 100 × 5 × 30 mm^3, were cut from the original larger stock. For each material batch, one beam specimen had its length along the rolling (L) direction and the other along the long transverse (LT). The IBRS component acting along the length of each beam specimen was measured using the slitting method, which comprises cutting the sample by wire EDM as shown in Fig. 1a (lower portion) in increments of cut depth, measuring strain after each cut depth and computing residual stress from the strain versus cut depth data [21]. Slitting provides a profile of the stress component perpendicular to the slitting plane (i.e., along the beam length in these experiments) as a function of distance along the slitting direction; the stress so determined reflects an average through the part depth (here along the 5 mm dimension of the beam). Strains were measured using metallic foil gages with active length 1.57 mm that were mounted on the back face of the beam specimens, opposite the start of the cut (see Fig. 1). Strains were measured after each of 40 equal cut depth increments, the final cut depth being equal to 95% of the 30 mm beam width. Residual stresses were computed using pulse regularization, as described earlier [21]. The residual stress measurement procedure for batch A was quite similar but for a larger plate thickness. The procedure and measurement results are presented in [22]. Measured IBRS are compared with literature values from Prime and Hill [21] (see section "Results") and used as an input for the FEM distortion model (see section "FEM Distortion Model").

Distortion Experiments

To validate the FEM distortion model, different samples, varying in their part geometry, IBRS, and milling path strategy, were manufactured. Besides, the position of the final part within the original block, here called "z-offset," was also varied. Two part geometries, varying in size and complexity, were machined: a simple small rib-typed structural part with two pocket

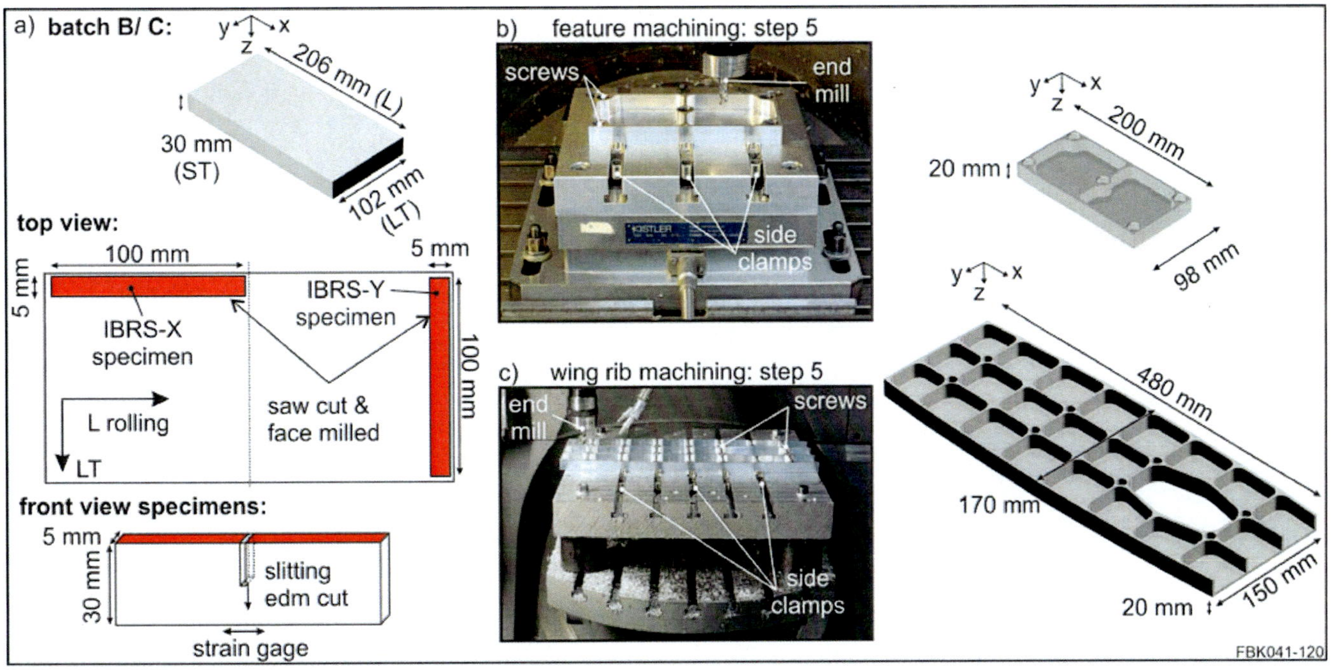

Fig. 1 Orientation of samples with slitting measurement plan (**a**) and experimental setups for machining of feature samples (**b**) and wing ribs (**c**)

features and one stiffener in the center (here called "feature sample," dimension 200 × 98 × 20 mm^3; see Fig. 1b) and a more complex and bigger geometry imitating a small-scaled airplane wing rib with multiple pockets and stiffeners (here called "wing rib," dimension 480 × 170 × 20 mm^3; see Fig. 1c) were analyzed. The feature sample was machined out of material batch B, originating from the same raw material, whereas the wing rib corresponds to material batch C. All components were machined on a five-axis DMG Mori DMU 70 CNC[1] machine. The manufacturing steps of the feature sample can be found in the literature [19]. The machining of the wing rib was similar (initial block size, 485 × 172 × 30 mm^3), but the order and clamping strategy deviated slightly (see Fig. 1c):

- Step 1: Face milling of backside in side clamps (feed per tooth $f_z = 0.2$ mm, cutting speed $v_c = 730$ m/min, radial cut depth $a_e = 40$ mm, axial cut depth $a_p = 1.5$ mm)
- Step 2: Side milling in side clamps ($f_z = 0.055$ mm, $v_c = 450$ m/min, $a_e = 2.5$ / 0.5 mm, $a_p = 4.4$ / 22 mm)
- Step 3: Drilling holes in side clamps
- Step 4: Face milling of top in side clamps and screws ($f_z = 0.2$ mm, $v_c = 730$ m/min, $a_e = 40$ mm, $a_p = 1.5$ mm)
- Step 5: Milling of pockets in side clamps and screws ($f_z = 0.2$ mm, $v_c = 200$ m/min, $a_e = 4$ mm, $a_p = 3$ mm)

A cutter with indexable inserts (Sandvik[1] R590-110504H-NL H10) was used to face mill the top and backside surface (step 1, 4) and a regular end mill (Kennametal[1] F3AA1200AWL) to side mill the walls (step 2) and pockets (step 5). MIRS from the cutting operations face milling (step 1, 4) and pocket milling (step 5) were measured in the previous research and plotted in Fig. 2. For more information we refer to [17–20]. Since the total machined part is 20 mm thick, different z-offsets of the part in the 30 mm stock are possible: a symmetrical position, removing 5 mm each at top and bottom, and an asymmetrical one, removing 1.5 mm at the bottom and 7.5 at the top were investigated. To achieve high feed rates, the pockets were milled in multiple layers with constant cutting parameters. The milling path strategy was varied from part to part from zig (machining in horizontal lines) to spiral path (machining from the inside out in spirals following the contour of the pockets). To prevent the distortion of the parts during pocket machining, the parts were held down by screws in addition to the clamping with side clamps. The backside surface was measured after the part was released from its clamping after step 5 with the coordinate measuring machine Tesa micro Hite 3D DCC[1] (repeatability limit ISO MPE-p 3.5 μm) to determine the distortion – here defined as the out of plane displacement in z-direction – caused by the RS. For the feature sample (wing rib), a spacing of the measured points of 2 mm (4 mm) with a distance to the edge of 1 mm (2 mm) was chosen, which resulted in 4714 (4419) measurement points in total. The final part distortion was analyzed by first leveling the data

[1] Naming of specific manufacturers is done solely for the sake of completeness and does not necessarily imply an endorsement of the named companies nor that the products are necessarily the best for the purpose.

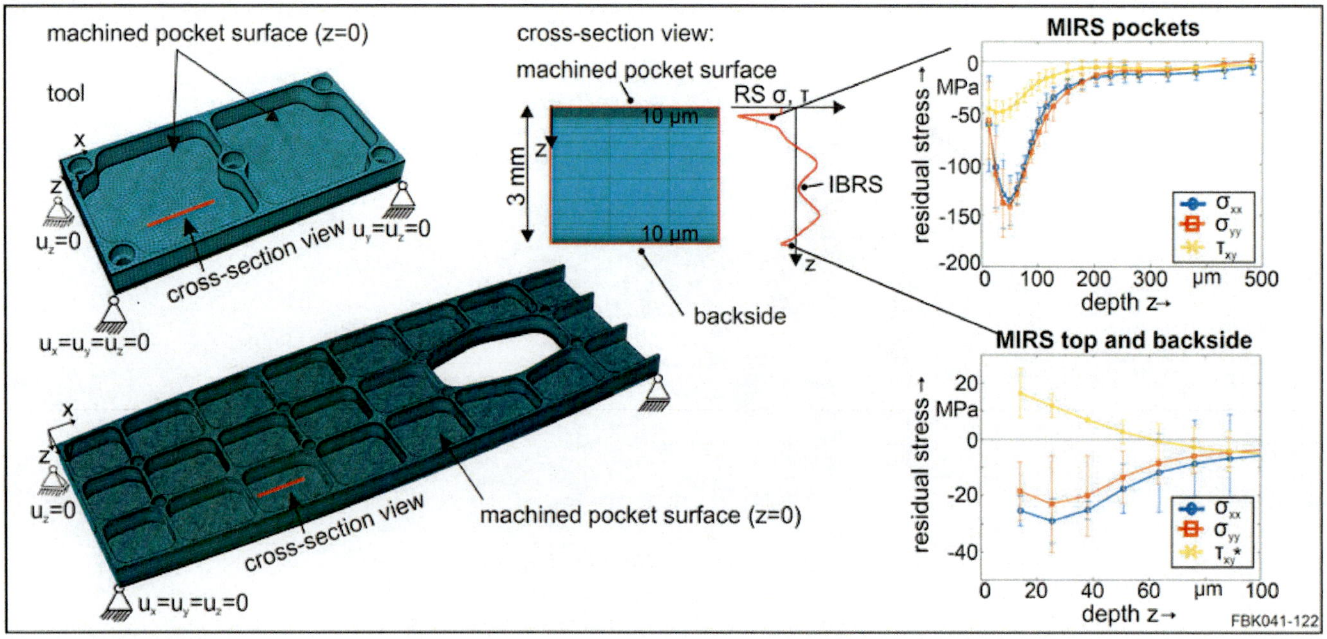

Fig. 2 FEM distortion model for feature sample and wing rib with MIRS from [23]

set (fitting a flat plane and subtracting it) and then setting a zero height at the resulting minimum, which was determined by averaging of the 10 (5) lowest measured points.

FEM Distortion Model

The developed FEM distortion model was presented and validated in previous research [17–20] and is now used to investigate the effect of varying IBRS on distortion for different part geometries. It is summarized here as follows: a static, lincar elastic FEM model was set up in ABAQUS[1] ($E = 71{,}700$ MPa, $\nu = 0.33$). The measured MIRS and IBRS were implemented as an initial condition (type = stress) to the final part shape without modelling the material removal process. The MIRS were limited to a near surface layer (pocket floor, 200 µm; backside/top, 70 µm; see Fig. 3). IBRS were applied to the rest of the bulk. All RS $\sigma(z)$ were linearly interpolated over depth z at the element centroids. For the spiral milling strategy, a coordinate transformation of the MIRS according to the direction of milling was achieved (for more information see [20]). Previous research showed that MIRS in the walls have neglectable impact on the distortion in z-direction [20] and were therefore not considered in the distortion model. The mesh of the feature sample (wing rib) consisted of 670,308 (2,096,508) eight-node brick (C3D8) elements. The global mesh size was 1.5 mm and was refined in the near surface layer down to 10 µm to resolve the MIRS properly. The parts were minimally constrained (3-2-1 constraint principle [15]) to avoid rigid body motion but to enable a free part distortion (see Fig. 2). After one simulation step (reaching RS equilibrium), the displacement at the parts backside was analyzed according to experiments (leveling and shifting data) and compared to the measured distortion.

Results

The results section is divided into two parts: the IBRS results and the distortion results including measurements and simulations.

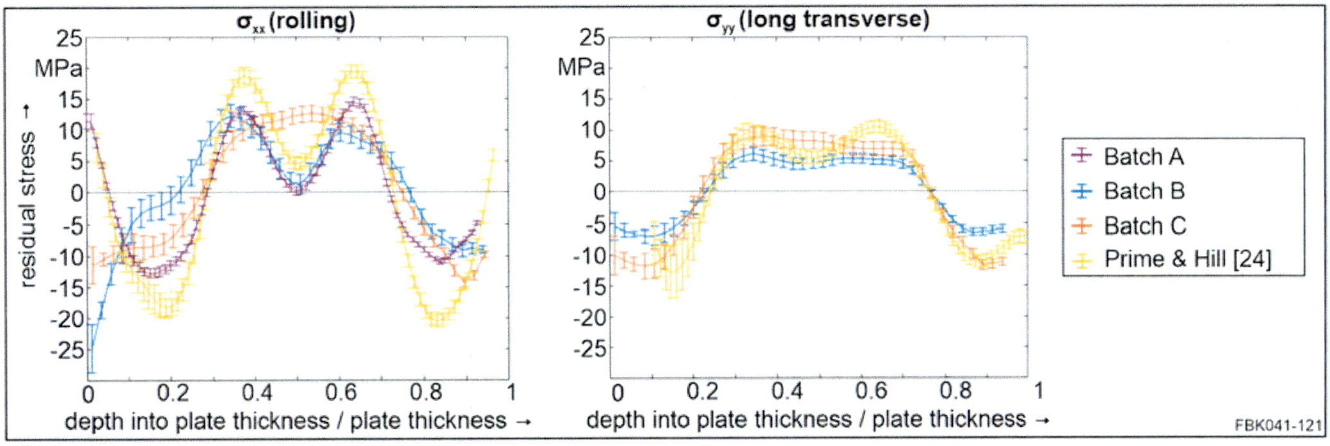

Fig. 3 Initial bulk residual stress comparison of measured and literature values of AA7050-T7451

Initial Bulk Residual Stress Measurements

Figure 3 depicts the measured IBRS and literature values from Prime and Hill [24] in x- and y-direction over the beam thickness normalized by the plate thickness. Most RS depth profiles showed the typical M shape of stresses with compressive RS close to the surface and tensile RS in the center. Batch C showed a ∩ shape rather than an M shape, which could be due to the distribution of strength through the thickness of the corresponding plate (see [24]). The peak magnitudes of IBRS in the three batches were similar, being 10–15 MPa, with stress magnitudes slightly larger along the rolling (x) direction than along the long transverse (y). The stress magnitudes in the present samples are smaller than those from the literature, which reached peak levels of 20 MPa. The trend in σ_{xx} for batch B at depths less than 20% of the plate thickness reflects outlying behavior, the cause for which is unclear. Overall, the results show residual stress with a clear, symmetric M shape in batch A, a less clear and less symmetric M shape in batch B, and a ∩ shape in batch C. The results are reasonably consistent with the prior work [24], which found similar trends in two plate thicknesses (80 mm and 25 mm), the thicker plate having a more distinct M shape.

Distortion

Figure 4 depicts the color maps of the measured distortion of the feature samples with different machining paths and z-offsets, as well as their simulated counterparts with different IBRS used as input. The orientation of the color maps is top view, meaning looking down to the milled pocket surface with positive distortion into the surface in positive z-direction. First the measurements are discussed before comparing simulated and measured distortions.

A slightly twisted distortion with the minima at the bottom left and top right corner and their maxima close to the top left and bottom right corner ($z_{max} = 0.211$ mm) was evident for the feature sample milled with zig milling strategy (see Fig. 4a). The shear MIRS induced a torsional moment in addition to the bending moment of the IBRS and the normal compressive MIRS [17–20, 23]. Changing the milling direction also resulted in a change of the distortion shape: a U shape with the maximum distortion in the middle of the part's length ($x = 100$ mm) and regions where the tool cut first in the material (highlighted with black arrows in Fig. 4) was evident ($z_{max} = 0.134$ mm). Due to the change of direction of the milling path to a spiral path, the distortion was reduced compared to the zig milling path, because the sign of the shear MIRS changed in 90° rotated regions. This led to an equilibration of the shear MIRS [20] and to a shift of the RS type dominating from MIRS to IBRS [23]. Choosing a z-offset of 5 mm of the final part in the raw material (Fig. 4c) instead of 1.5 mm decreased the distortion ($z_{max} = 0.104$ mm), because the IBRS were then symmetrically distributed over the height z, inducing therefore a smaller bending moment [23].

The shape, including the position of the maximum distortion, was predicted correctly by all simulations using different IBRS inputs – except for regions where the tool ramped in for the spiral milling path (highlighted with black arrows in Fig. 4). This could be attributed to the presence of deviating MIRS in those regions due to different machining kinematics. Surprisingly, the level of distortion varied significantly when using different IBRS as input for all investigated cases, although IBRS magnitude varied only by about 5 MPa. The distortion level was predicted best with the simulations using the IBRS

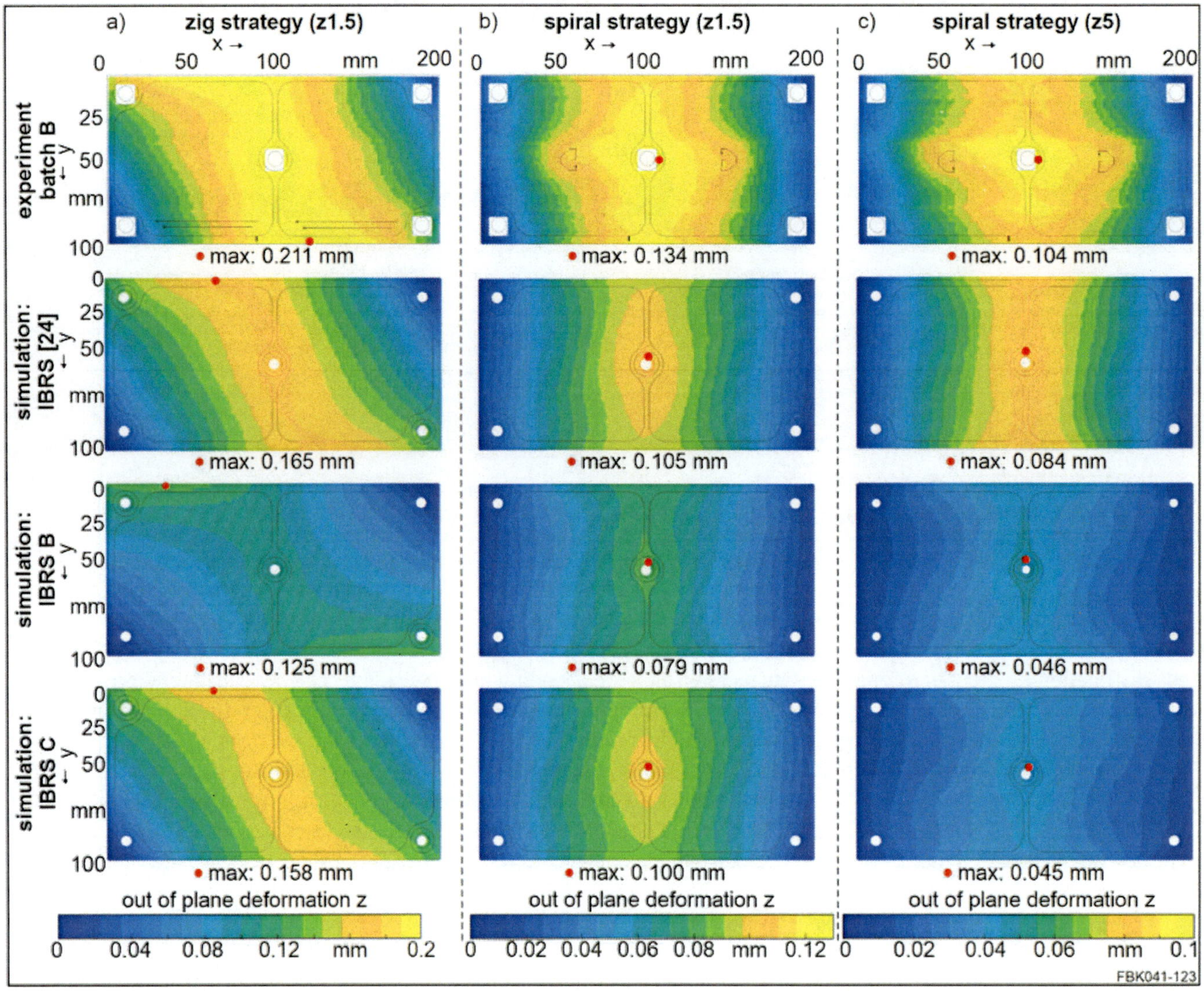

Fig. 4 Color maps of the measured and simulated distortions of the feature sample for different milling paths, z-positions and IBRS: zig strategy z1.5 (**a**), spiral strategy z1.5 (**b**) and spiral strategy z5 (**c**)

literature values from Prime and Hill [24] instead of the measurements done on samples from the same batch B material. Deviations (in terms of differences between prediction and measurement) of the maximum distortion down to 22% (Fig. 4a; zig – z1.5), 21% (Fig. 5b; spiral – z1.5), and 19% (Fig. 4c; spiral – z5) were found. All trends discussed for the measured distortions were matched by the simulations.

Figure 5 highlights the color maps of the measured distortion of the wing rib with different machining paths, as well as their simulated counterparts with different IBRS used as input. For the distortion measurements, the same trends as discussed for the smaller feature sample were found: a twisted distortion shape for zig milling (z_{max} = 0.739 mm) and a U-shaped reduced distortion for spiral milling (z_{max} = 0.676 mm) were evident.

The wing rib shape was predicted correctly by all simulations using different IBRS inputs. Small deviations for the position of the maximum distortion for zig milling strategy were found. This could be attributed to the fact that in experiments outer machining paths were not perfectly horizontal as assumed in the simulation, due to the curved shape of the part. Again, the highest accuracy of the simulation was reached when using the IBRS from the literature as input. Deviations down to 7% (Fig. 5a; zig – z5) and 28% (Fig. 5b; spiral – z5) were achieved. Using the measured IBRS led to an underestimation of the distortion for both investigated cases. One possible explanation for this trend could be that the IBRS in the beam specimens deviate from the IBRS in the remainder of the plate stock (which is perhaps closer to the literature values); this could occur if the beams were closer to the edges of the original plate stock, but this remains unknown.

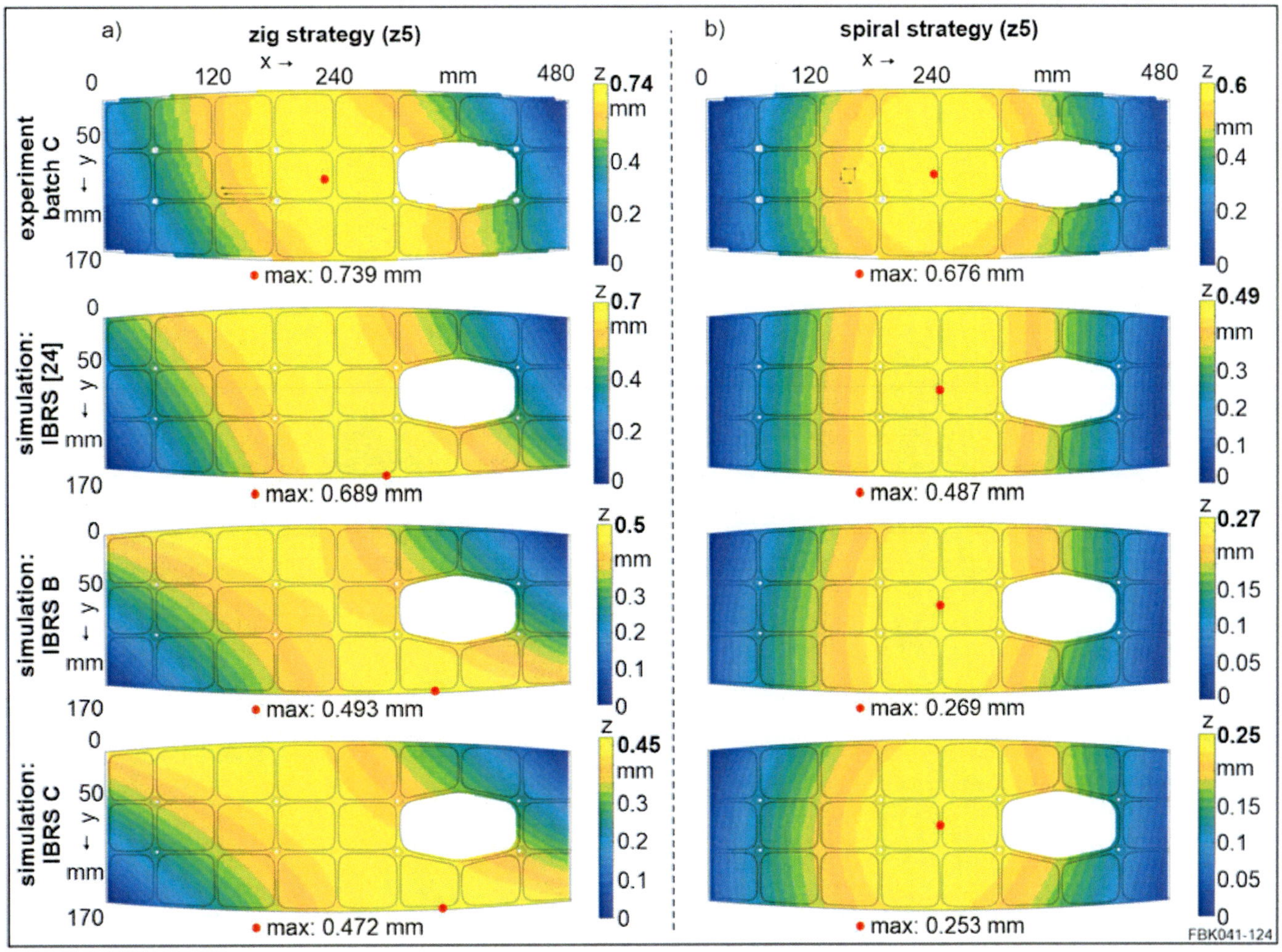

Fig. 5 Color maps of the measured and simulated distortions of the wing rib for different milling paths and IBRS: zig strategy z5 (**a**) and spiral strategy z5 (**b**)

Conclusion

IBRS measurements of different batches of A7050-T7451 are reasonably consistent with what was found in the literature by Prime and Hill [24] with a clear, symmetric M shape in batch A, a less clear and less symmetric M shape in batch B, and a ∩ shape in batch C. The magnitude of the measured IBRS was slightly smaller than literature values (by about ~5 MPa). When using the different IBRS as input for the FEM distortion model in addition to the MIRS, the shape of distortion was predicted correctly for different geometries and milling strategies and for different z-offsets. However, the level of distortion, identified by the maximum, was underestimated when using the measured IBRS as input. The best results were achieved when using the IBRS literature values as input, in which case deviations of the maximum distortion of 28% (spiral-strategy) and 7% (zig-strategy) were evident.

Acknowledgments The authors would like to thank the German Research Foundation (DFG, Germany) and the National Science Foundation (NSF, USA) for the financial support within the project AU 185/64-1 (351381681) "NSF DFG Collaboration to Understand the Prime Factors Driving Distortion in Milled Aluminum Workpieces" (NSF funding Award No. 1663341) and within the International Research Training Group 2057 – Physical Modeling for Virtual Manufacturing (IRTG2057), Funding No. 252408385. Any opinions, findings, and conclusions or recommendations expressed in this material are those of the authors and do not necessarily reflect the views of the NSF or DFG.

References

1. Klocke, F.: Fertigungsverfahren 1 – Zerspanung mit geometrisch bestimmter Schneide. Springer Vieweg, Berlin (2018)
2. Jayanti, S., Ren, D., Erickson, E., Usui, S., Marusich, T., Marusich, K., Elanvogan, H.: Predictive modeling for tool deflection and part distortion of large machined components. In: Procedia CIRP 12 – Proceedings of the 8th CIRP Conference on Intelligent Computation in Manuf. Eng, pp. 37–42, Naples (2013)
3. Berky, E.: Aerostructures made in Augsburg. In: International ARO Seminar Aluminium-HSC Machining, Reutte (A) (2001)
4. Li, J.G., Wang, S.Q.: Distortion caused by residual stresses in machining aeronautical aluminum alloy parts: recent advances. Int. J. Adv. Manuf. Technol. **89**, 997–1012 (2016)
5. Denkena, B., Boehnke, D., de Leon, L.: Machining induced residual stress in structural aluminum parts. Prod. Eng. **2**, 247–253 (2008)
6. Perez, I., Madariaga, A., Cuesta, M., Gray, A., Arrazola, P.J., Ruiz, J.J., Rubio, F.J., Sanchez, R.: Effect of cutting speed on the surface integrity of face milled 7050-T7451 aluminium workpieces. In: Procedia CIRP 71 – Proceedings to the 4th CIRP Conference on Surface Integrity, pp. 460–465. Elsevier BV (2018)
7. Rao, B., Shin, Y.C.: Analysis on high-speed face-milling of 7075-T6 aluminum using carbide and diamond cutters. Int. J. Mach. Tools Manuf. **41**(12), 1763–1781 (2001)
8. Tang, Z.T., Liu, Z.Q., Wan, Y., Ai, X.: Study on residual stresses in milling aluminium alloy 7050-T7451. In: Yan, X.-T., Jiang, C., B. (eds.) Advanced Design and Manufacture to Gain Competitive Edge: New Manufacturing Techniques and Their Role in Improving Enterprise Performance, pp. 169–178. Springer, London (2008)
9. Brooks, C.R.: Principles of heat treating of nonferrous alloys: heat treating of aluminum alloys. In: ASM Handbook 4 Heat Treating, pp. 1861–1960. ASM International (1991)
10. Keleshian, N., Kyser, R., Rodriquez, J.: On the distortion and warpage of 7249 aluminum alloy after quenching and machining. J. Mater. Eng. Perform. **20**, 1230–1234 (2011)
11. Chantzis, D., Van-der-Veen, S., Zettler, J., Sim, W.M.: An industrial workflow to minimise part distortion for machining of large monolithic components in aerospace industry. In: Procedia CIRP 8 – Proceedings of 14th CIRP Conference on Modeling of Machining Operations, pp. 281–286. Elsevier (2013)
12. Zhang, Z., Li, L., Yang, Y., He, N., Zhao, W.: Machining distortion minimization for the manufacturing of aeronautical structure. Int. J. Adv. Manuf. Technol. **73**(9), 1765–1773 (2014)
13. Cerutti, X., Mocellin, K.: Influence of the machining sequence on the residual stress redistribution and machining quality: analysis and improvement using numerical simulations. Int. J. Adv. Manuf. Technol. **83**, 489–503 (2016)
14. Dreier, S.: Simulation des eigenspannungsbedingten Bauteilverzugs, Dissertation, Hannover (2018)
15. Huang, X.M., Sun, J., Li, J.F.: Finite element simulation and experimental investigation on residual stress-related monolithic component deformation. Int. J. Adv. Manuf. Technol. **77**, 1035–1041 (2015)
16. Madariaga, A., Perez, I., Arrazola, P.J., Sanchez, R., Ruiz, J.J., Rubio, F.J.: Reduction of distortions in large aluminum parts by controlling machining-induced residual stresses. Int. J. Adv. Manuf. Technol. **97**, 967–978 (2018)
17. Weber, D., Kirsch, B., Chighizola, C.R., D'Elia, C.R., Linke, B.S., Hill, M.R., Aurich, J.C.: Analysis of machining induced residual stresses of milled aluminum workpieces, their repeatability and their resulting distortion. Int. J. Adv. Manuf. Technol. **115**, 1089–1110 (2021)
18. Chighizola, C.R., D'Elia, C.R., Weber, D., Kirsch, B., Aurich, J.C., Linke, B.S., Hill, M.R.: Intermethod comparison and evaluation of near surface residual stress in milled aluminum. Exp. Mech. **61**, 1309–1322 (2021)
19. Weber, D., Kirsch, B., Chighizola, C.R., D'Elia, C.R., Linke, B.S., Hill, M.R., Aurich, J.C.: Investigation on the scale effects of initial bulk and machining induced residual stresses of thin walled milled monolithic aluminum workpieces on part distortions: experiments and finite element prediction model. In: Procedia CIRP 102 – Proceedings of the 18th CIRP Conf. on Modeling of Machining Operations, pp. 337–342. Elsevier (2021)
20. Weber, D., Kirsch, B., Jonsson, J.E., D'Elia, C.R., Linke, B.S., Hill, M.R., Aurich, J.C.: Simulation based compensation techniques to minimize distortion of thin-walled monolithic aluminum parts due to residual stresses. CIRP J. Manuf. Sci. Technol. **38**, 427–441 (2022)
21. Hill, M.R.: The slitting method. In: Schajer, G.S. (ed.) Practical Residual Stress Measurement Methods. Wiley, West Sussex (2013)
22. Chighizola, C.R., Hill, M.R.: Two-dimensional mapping of bulk residual stress using cut mouth opening displacement. Exp. Mech. **62**, 75–86 (2022)
23. Weber, D., Kirsch, B., D'Elia, C.R., Jonsson, J.E., Linke, B.S., Hill, M.R., Aurich, J.C.: Simulation based investigation on the effect of the topology and size of milled thin-walled monolithic aluminum parts on the distortion due to residual stresses. In: Proceedings of the International Conference on Residual Stresses (ICRS) 11, Archive ouverte HAL hal-03869964, Nancy (2022)
24. Prime, M.B., Hill, M.R.: Residual stress, stress relief, and inhomogeneity in aluminum plate. Scr. Mater. **46**(1), 77–82 (2002)

On the Use of iDIC (Integrated Digital Image Correlation) for the Slitting Method

P. M. Santucci, A. Baldi, D. Lai, and G. Marongiu

Abstract The slitting method is a well-known technique for residual stress measurement. It consists of cutting a slot in the specimen while monitoring the relieved strains with one or more strain gauges. The stress distribution is then computed by solving a reverse problem.

The use of Digital Image Correlation to replace the standard strain-gauge-based approach has been proven successful in previous work. In this new contribution, the authors want to explore the integrated Digital Image Correlation (iDIC) potential as a direct replacement for the entire procedure. Instead of using DIC as an optical extensometer—i.e., using a large DIC subset to extract single-point data—in the iDIC formulation, the shape functions describing the surface displacement field satisfy the equilibrium conditions. Thus, the minimization parameters (both in space and in time) are the Legendre polynomial weights directly connected to the residual stress values.

To verify the new approach, an aluminum beam is loaded in bending above the yield stress and then unloaded using a four-point configuration; then, the slot is performed using a milling machine while imaging the back face. The use of strain gauges allows comparing the results of the new approach with the standard one.

Keywords Residual stress · Slitting · Digital Image Correlation · iDIC

Introduction

Slitting is one of the most popular destructive approaches for evaluating the residual stress distribution throughout a material sample. Historically, the first use of such technique is assessed in the article by Schwaighofer in 1964 [1]. In the following decades, the technique has been studied and improved until today [2–4].

Like hole drilling, slitting uses the removal of material from the specimen to relieve residual stresses [4, 5]. The material is removed along a linear path leading to the cutting of a slit, hence the name slitting. The internal relaxation of the residual stress is measured using a strain gauge bonded to the back of the specimen, usually aligned with the slit axis [4, 5]. The problem can be approached from two different perspectives, a direct and an inverse method. The direct approach uses the known residual stress distribution to evaluate the strain on the surface as the slit depth increases. However, the inverse problem consists of estimating the residual stress distribution using the strain measurement on the surface [4]. Considering the nature of the residual stress, the inverse method seems to be more suitable for the industrial reality where the residual stress field is unknown.

In the inverse approach, the unknown residual stress distribution is defined as series that could be polynomial, harmonic, or a power series [4]. The Legendre polynomial basis is widely used in literature:

$$\sigma_{yy}(x) = \sum\nolimits_{j}^{m} A_j P_j(x) = [P]\{A\} \tag{1}$$

The series in Eq. (1) expresses the residual stress as a function of depth (see Fig. 1b for coordinate system), in which P_j is the Legendre basis functional and A_j is the weights. The order of the series m usually lies between 4 and 12, and its value depends on error minimization considerations [6].

P. M. Santucci · A. Baldi (✉) · D. Lai · G. Marongiu
Department of Mechanical, Chemical and Materials Engineering, University of Cagliari, Cagliari, Italy
e-mail: antonio.baldi@dimcm.unica.it

C. Franck et al. (eds.), *Challenges in Mechanics of Biological Systems and Materials, Thermomechanics and Infrared Imaging, Time Dependent Materials and Residual Stress, Volume 2*, Conference Proceedings of the Society for Experimental Mechanics Series,
https://doi.org/10.1007/978-3-031-50470-9_6

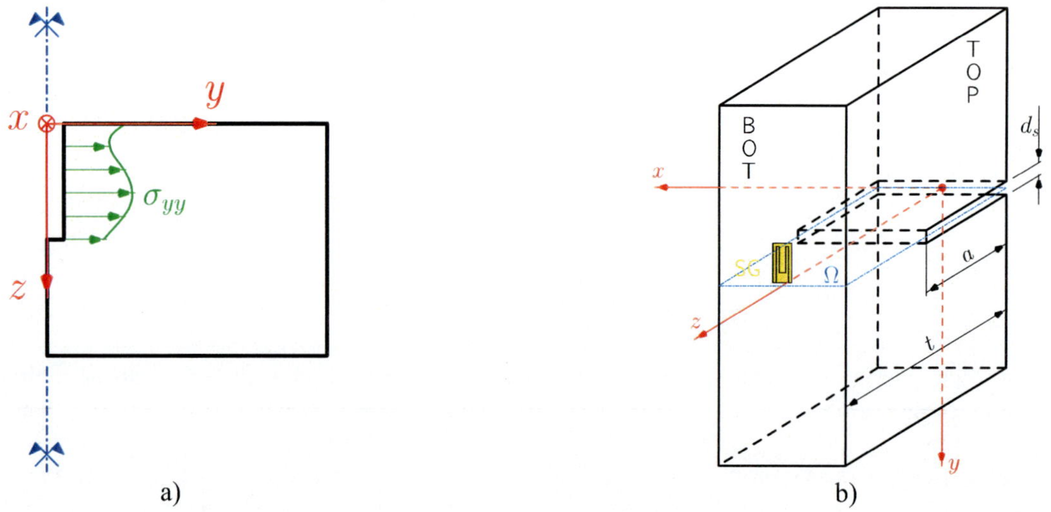

Fig. 1 Model geometry. (**a**) Coordinate system in 2D scheme. (**b**) 3D geometry of specimen

The weights A_j are the final target of our inverse approach, which consists of measuring the strain on the sample surface. Under the hypothesis of a linear elastic material model, the strain field through depth depends on the compliance matrix $[C]$. Each term of $[C]$ is the strain read by the strain gauge at its location, due to a residual stress expressed as a Legendre function for a given slit depth [4, 5]. The terms in the $[C]$ matrix are computed using finite element simulations with consistent geometric parameters [4, 5, 7].

$$\varepsilon_{yy}(x) = \sum_j^m A_j C_j(x) = [C]\{A\} \tag{2}$$

Equation (2) outlines the strain field through depth using the superposition principle [4, 5]. The use of a least square approach gives the solution for our problem:

$$\{A\} = \left([C]^{\mathrm{T}}[C]\right)^{-1}[C]^{\mathrm{T}}\left\{\tilde{\varepsilon}\right\} \tag{3}$$

where $\left\{\tilde{\varepsilon}\right\}$ is a vector containing the measured strain for each depth stage of the slit. Once the weights $\{A\}$ of the Legendre function are known, the stress distribution can be computed via back-substitution in Eq. (1).

The use of strain gauges is assessed in a wide literature overview for its robustness and because the slit is often cut via EDM: in this case the specimen is immersed under water, and a suitably coated strain gauge is the best solution for measuring the strain.

In this article we propose the Digital Image Correlation (DIC) to measure the strain on the sample surface. In the literature there are few articles on the use of DIC with slitting [8–10], while this technique has been used with the hole drilling approach in different works [9–14].

The DIC poses its basis on the computer vision field and more precisely on image matching technique [15]. The basic hypothesis of DIC is the constancy of the optical flow between a reference image I and a target image G, defined as:

$$\frac{\partial I}{\partial t} + \frac{\partial I}{\partial x}\frac{\partial u}{\partial t} + \frac{\partial I}{\partial y}\frac{\partial v}{\partial t} = 0 \tag{4}$$

where $I = I(x, y, t)$ is the intensity field at time t while u and v are the x and y components of the displacement. The problem is clearly ill-posed because (2) involves two unknown displacement components and only one equation is available. Various strategies have been proposed in the literature to solve this problem [16, 17], but the most widely used approach is the Lukas-Kanade [18].

This approach consists of partitioning the image domain in smaller regions called subset. Thus, the displacement components are expressed as shape functions around a point.

$$u = p_0 + p_1\xi + p_2\eta + p_3\xi^2 + p_4\xi\eta \tag{5a}$$

$$v = q_0 + q_1\xi + q_2\eta + q_3\xi^2 + q_4\xi\eta \tag{5b}$$

where p_i and q_j are the parameters of the shape functions while ξ and η are the axes of a local coordinate system centered on the subset.

This reduces the number of unknowns from twice the number of subset pixels to the shape function parameters.

$$\chi^2 = \sum_r \sum_c [I\ (r, c) - G\ (r + u, c + v)]^2 \tag{6}$$

Equation (6) defines a cost function χ^2 that leads to a local least square minimization. The summations in Eq. (6) are extended over the pixels of the subset, following the r counter for the rows and the c counter for the columns.

The shape functions presented in Eqs. (5a) and (5b) are general purpose. Some problems have a displacement field that can be expressed analytically: it is therefore more convenient to use specific shape functions that define the displacement field globally [14]. This procedure is called integrated Digital Image Correlation (iDIC). The slitting method uses a series to model the strain field due to the relaxation of the residual stress (i.e., Eq. (2)); so, the displacement can be defined using the same polynomial series.

$$v_i(y) = \sum_j^n A_j D_{ij}(y) \tag{7}$$

Equation (7) defines the displacement component along the slit edge, identified by the z axis in Fig. 1a. The displacement field is defined at the i-th stage of cutting depth: D_{ij} is the polynomial basis for the i-th stage of cutting and j-th order of polynomial and A_j is the weight at j-th order.

$$v_i^{\mathrm{SF}}(y) = v_i + \sum_j^m A_j D_{ij}(y) \tag{8}$$

The shape function v_i^{SF} in Eq. (8) is defined using the model of Eq. (7) and adding a term for the rigid motion. The advantage of using iDIC is the significant amount of data in the bottom side (Fig. 1b) of the specimen: to make full use of the data, a shape function for the transverse component of the displacement can be developed by considering the Poisson effect.

$$\nu = -\frac{\varepsilon_x}{\varepsilon_y} = -\frac{\frac{\partial u}{\partial x}}{\frac{\partial v}{\partial y}} \tag{9}$$

By differentiating Eq. (8), substituting into Eq. (9), and integrating, we get:

$$u_i^{\mathrm{SF}}\ (x, y) = -\nu\, x \sum_j A_j \frac{\partial}{\partial y} D_{ij}(y) + u_0(y) \tag{10}$$

where $u_0(y)$ is the integration constant. Posing the reference system on the symmetry axis of the bottom face leads to a null value of $u_0(y)$. Under this assumption the shape functions for the iDIC model are:

$$u_i^{\mathrm{SF}}\ (x, y) = -\nu\, x \sum_j A_j \frac{\partial}{\partial y} D_{ij}(y) \tag{11a}$$

$$v_i^{SF}(y) = v_i + \sum_j^m A_j D_{ij}(y) \tag{11b}$$

Once the shape functions are defined, the solution follows the step of a classic DIC analysis.

Background

The experimental setup is developed using both the strain gauge bonded to the specimen surface and an Allied Vision F421b camera for DIC measures (with resolution of 2048 × 2048).

The specimen is a 6082-T6 aluminum beam with a rectangular cross-section (30 mm × 14.4 mm). The specimen was loaded using a four-point bending configuration to introduce a residual stress field. The residual surface strain has been measured using a strain gauge.

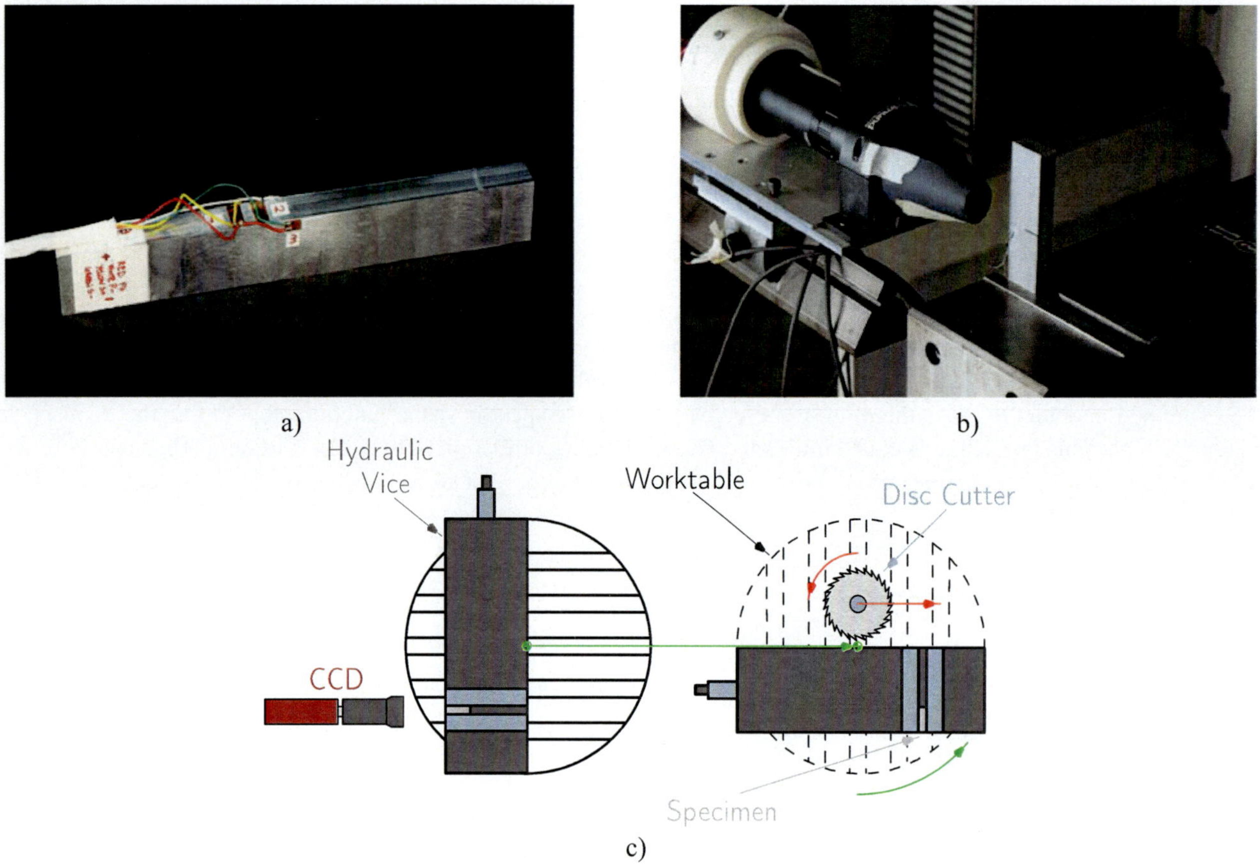

Fig. 2 Specimen and experimental setup. (**a**) Specimen overview with strain gauges. (**b**) CCD configuration. (**c**) Global scheme of setup

The strain gauges used in the test are arranged in a full bridge configuration: the two active strain gauges are bonded to the faces parallel to the $y - z$ plane, while the two dummies are bonded to the bottom face (Fig. 2a).

The test is carried out using a five-axis CNC machine, a Deckel Maho DMU 60 P hi-dyn. The specimen is held in position by a hydraulic vice mounted on the machine worktable (Fig. 2b). The vice moves between two positions (Fig. 2c):

- The cutting point, where the slit is cut step by step, which is far away from the CCD to avoid dust and dirt deposits on the lens
- The acquisition point, where the specimen presents the bottom side to the CCD after each cutting step

The experimental procedure begins with the removal of the layer of material on the slit. Prior to the test, the disc cutter is calibrated to ensure zero positioning on the top surface of the sample. After the material removal step, the sample is moved near the camera taking the bottom side toward it. There is a pause of 1 minute between positioning and acquisition of the two images and strain gage data.

Analysis

The analysis of results has been performed in parallel using the strain gauge data and DIC analyses on images.

Figure 3a, b shows the strain and stress distribution as the slit depth increases. The data were acquired with the full strain gauge bridge mounted on the back of the sample (Fig. 2a). This set of data is used as a reference to validate the DIC analysis.

The data acquired using the CCD are processed using a standard DIC analysis. The strains are estimated using a ROI defined over the strain gauge position in the sample side. The strains along the y axis are computed as a mean value over the width of the ROI for each slit depth. The results of DIC analysis are shown in Fig. 4.

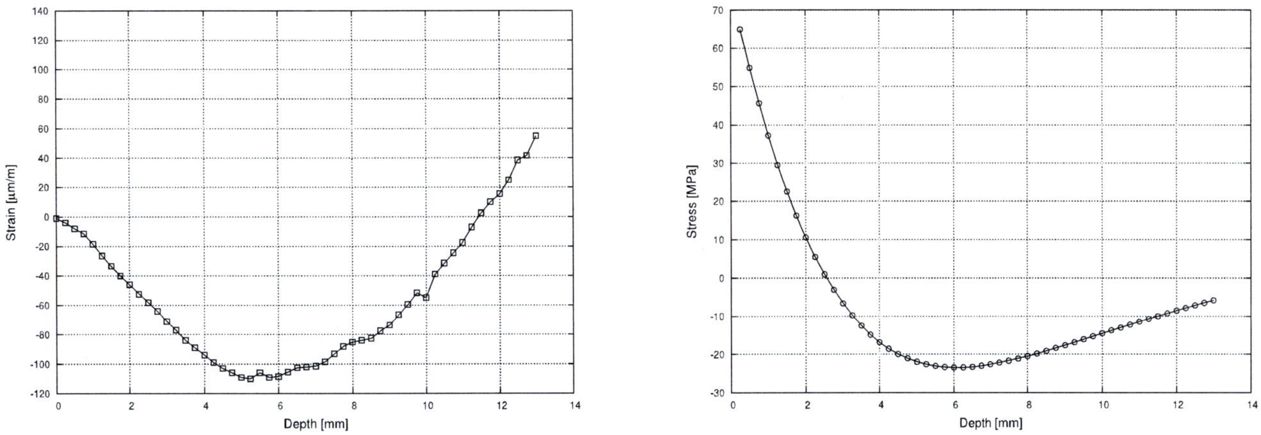

a) b)

Fig. 3 Experimental data acquired by the strain gauges. (**a**) Strain distribution through thickness. (**b**) Stress distribution through thickness

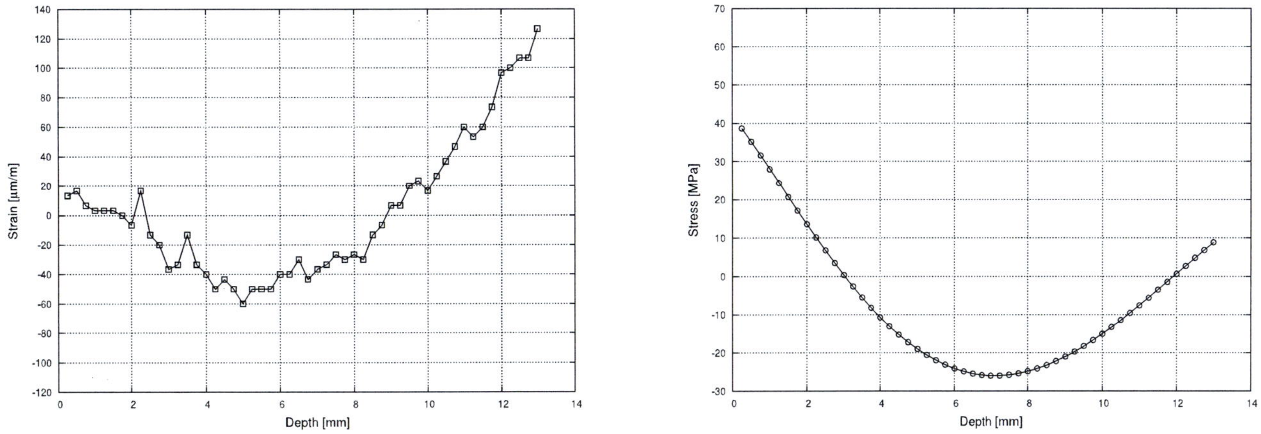

a) b)

Fig. 4 Experimental data from DIC analysis. (**a**) Strain field. (**b**) Stress field

Conclusion

The DIC analysis was performed using the ROI as a virtual strain gauge. The results shown in Fig. 4 are close to the reference ones, obtained using the classic approach proposed in literature. Note that strain values differ due to the different position of strain gauges in the specimen side compared to the ROI position in the back. The calibration coefficients, i.e., the weight of the Legendre polynomial, differ in the same fashion because they depend locally on the position on the sample surface, but the estimate stress curves are almost the same.

Based on these results, the iDIC analysis performed on the same dataset should increase the precision of results, because it involves a significantly larger number of data point, thus increasing the robustness of the method.

References

1. Schwaighofer, J.: Determination of residual stresses on the surface of structural parts. Exp. Mech. **4**(2), 54–56 (1964). https://doi.org/10.1007/BF02323965
2. Soete, W., Vancrombrugge, R.: Determination of residual stresses below the surface. In: Residual Stresses in Metals and Metal Construction, vol. 8, pp. 17–28. Reinhold Publishing Corporation, New York (1954)
3. Cheng, W., Finnie, I.: A method for measurement of axisymmetric axial residual stresses in circumferentially welded thin-walled cylinders. J. Eng. Mater. Technol. **107**(3), 181–185 (1985). https://doi.org/10.1115/1.3225799
4. Prime, M.B.: Residual Stress Measurement by Successive Extension of a Slot: a Literature Review. Los Alamos National Lab, New Mexico (1997)

5. Schajer, G.S.: Practical Residual Stress Measurement Methods. John Wiley & Sons, Chichester (2013)
6. Prime, M.B., Hill, M.R.: Uncertainty, model error, and order selection for series-expanded, residual-stress inverse solutions. J. Eng. Mater. Technol. **128**, 175 (2006)
7. Schajer, G.S.: Application of finite element calculations to residual stress measurements. J. Eng. Mater. Technol. **103**(2), 157–163 (1981). https://doi.org/10.1115/1.3224988
8. Salehi, S.D., Rastak, M.A., Shokrieh, M.M., Barrallier, L., Kubler, R.: Full-field measurement of residual stresses in composite materials using the incremental slitting and digital image correlation techniques (2020). https://doi.org/10.1007/s11340-020-00640-2
9. Xu, Y., Bao, R.: Residual stress determination in friction stir butt welded joints using a digital image correlation-aided slitting technique. Chin. J. Aeronaut. **30**(3), 1258–1269 (2017). https://doi.org/10.1016/j.cja.2016.11.003
10. Blair, A., et al.: Residual stress relaxation measurements across interfaces at macro-and micro-scales using slitting and DIC. J. Phys. Conf. Ser. **181**(1) (2009). https://doi.org/10.1088/1742-6596/181/1/012078
11. Nelson, D.V., Makino, A., Schmidt, T.: Residual stress determination using hole drilling and 3D image correlation. Exp. Mech. **46**(1), 31–38 (2006). https://doi.org/10.1007/s11340-006-5859-0
12. Nelson, D.V.: Residual stress determination by Hole drilling combined with optical methods. Exp. Mech. **50**(2), 145–158 (2010). https://doi.org/10.1007/s11340-009-9329-3
13. Lord, J.D., Penn, D., Whitehead, P.: The application of digital image correlation for measuring residual stress by incremental hole drilling. Appl. Mech. Mater. **13–14**, 65–73 (2008). https://doi.org/10.4028/www.scientific.net/AMM.13-14.65
14. Baldi, A.: Residual stress measurement using hole drilling and integrated digital image correlation techniques. Exp. Mech. **54**(3), 379–391 (2014). https://doi.org/10.1007/s11340-013-9814-6
15. Schreier, H., Orteu, J.-J., Sutton, M.A., et al.: Image Correlation for Shape, Motion and Deformation Measurements: Basic Concepts, Theory and Applications, vol. 1. Springer, New York (2009)
16. Horn, B.K.P., Schunck, B.G.: Determining optical flow. Artif. Intell. **17**, 185 (1981)
17. Bruhn, A., Weickert, J., Schnörr, C.: Lucas/Kanade meets Horn/Schunck: combining local and global optic flow methods. Int. J. Comput. Vis. **61**, 211 (2005)
18. Lucas, B.D., Kanade, T.: An iterative image registration technique with an application to stereo vision. In: Proc 7th Intl Joint Conf on Artificial Intelligence (IJ CAI). ACM (1981)

A Rapid Procedure to Reconstruct S/N Curve by Using Harmonic Components of Thermal Signal

R. De Finis, D. Palumbo, and U. Galietti

Abstract The use of thermal indexes assessed by monitoring rapid fatigue tests with infrared detectors as reliable damage parameters to reconstruct the S/N curve is currently a topic still presenting open points. First of all, the selection of the proper thermal index representing damage is a topic to be explored, and also the relationship between the damage from rapid stepwise tests and constant amplitude tests is another point of discussion.

In the present work, we deal with such an issue by investigating the first amplitude harmonic (FAH) of thermal signal related to thermoelastic phenomena and dissipative effects too. It has been demonstrated that FAH is related to stiffness degradation and stress-induced effects. Moreover, it provides a local analysis of specific effect related to the material fatigue damage without artefacts.

The results show that due to the relationship between stiffness degradation and FAH and specific material properties, it is possible to reconstruct the S/N curve by carrying out just one constant amplitude test and a stepwise rapid test. Moreover, the capability of temperature FAH to study fatigue behaviour and detect damage during any loading procedure is also presented.

Keywords S/N curve · Stiffness degradation · Stepwise loading · Temperature first amplitude harmonic · Thermoelastic signal

Introduction

The fatigue behaviour of composites is crucial for the structural analysis due to damage mechanisms that vary depending on layup, mechanical properties and technological process of constituents. As well known, the damage can be understood as mechanical property degradation that can occur in different stages leading to final material failure [1–6]. In this way, the damage analysis is essential to develop 'stiffness-driven' design procedures that are reliable to implement a damage-tolerant approach.

Several models for damage analysis have been presented in the literature [1, 5, 6]; each one is tailored on the specific composite and application (e.g. woven [7], FRP [6], etc.) and any type of damage [1]. However, to validate these models, classic experimental tests are required which involve experimental campaigns and analyses that are time-consuming and require a lot of resources for modelling and validation stages. These implications clearly affect the time-to-market of new components and materials. This problem has, in recent years, become the 'driving force' for the development of new test procedures for rapid and effective fatigue test campaigns, supported by experimental techniques such as infrared thermography [6–18].

Nowadays, the stepwise loading procedure, to induce a self-heating in the material, can provide an estimation of the material endurance limit using the thermographic method on a small number of specimens [14]. Even if the approach proved to be effective on metallic and non-metallic materials [6–8], as far as composites are concerned, the relationship between the thermal parameters from a stepwise test and those from constant amplitude tests is still a topic to be explored. Furthermore, the choice and evaluation of a correct damage parameter that takes into account the specific accumulation mechanisms during

R. De Finis (✉)
Department of Engineering for Innovation, University of Salento, Lecce, Italy
e-mail: rosa.definis@unisalento.it

D. Palumbo · U. Galietti
Department of Mechanics, Mathematics and Management, Polytechnic University of Bari, Bari, Italy

C. Franck et al. (eds.), *Challenges in Mechanics of Biological Systems and Materials, Thermomechanics and Infrared Imaging, Time Dependent Materials and Residual Stress, Volume 2*, Conference Proceedings of the Society for Experimental Mechanics Series,
https://doi.org/10.1007/978-3-031-50470-9_7

the rapid fatigue test are also a topic to be explored. Huang et al. [6], for example, proposed a damage parameter based on the stiffness degradation of composite specimens and the average temperature measured during rapid fatigue tests. The procedure required the assessment of material constants from the classic fatigue test campaign; therefore, it was in any case challenging from the point of view of tests and analyses. Montesano et al. [8] proposed a method based on the temperature continuously acquired during the steady-state stage of a loading block of stepwise fatigue tests, accounting for the procedure proposed by La Rosa and Risitano [14] on metals. The basic condition for the application of the procedure is the achievement of thermal steady state in the temperature evolution and also the hypothesis of energy conservation (constant energy parameter) which should be verified for composites. In this regard, in the last 10 years, more effective indices have been studied, such as the harmonic components of the temperature signal [12, 16], to carry out the analysis of the damage processes in composites.

In the works of De Finis et al. [12], the relationship between the mechanical work associated with the material (area under the hysteresis cycle) and the thermal energy dissipated during tests of constant amplitude was presented, and, in another work, [17], the ability of the thermoelastic signal to describe the degradation of the material stiffness was investigated. These indices have proved to be very effective for estimating the residual life of the composites, as they are directly linked to the damage.

The aim of present work is to obtain the fatigue life of a quasi-isotropic composite via the assessment of thermal indices acquired during a rapid fatigue test.

The novelties and outputs of present research are related to:

- Capability of temperature first amplitude harmonic (FAH) to detect fatigue damage
- Capability of temperature FAH to study fatigue behaviour
- Correlation between FAH and stiffness degradation during any loading condition
- An innovative procedure to reconstruct the S/N curve from FAH data acquired during a stepwise loading

Background

Damage and Stiffness Loss of Material

As discussed by several authors [1–3], for a composite subjected to uniaxial fatigue testing (constant amplitude), the actual damage parameter related to the i-th stress level, D_i, is:

$$D_i = 1 - \left(\frac{E}{E_0}\right)_i \tag{1}$$

where D_i of course varies also depending on loading cycles. During a cyclic loading test, the first damage mechanism that occurs is the appearance of transverse cracks in the layers where fibres are oriented in a direction different from the one of the loading [3, 4, 6]. From Eq. (1), it results that D_i can be defined as the Young's modulus loss (the actual value of longitudinal one in this specific case, E_i is referred to a reference value usually measured at the beginning of the test or evaluated from the static tensile tests, E_0). The stiffness degradation is related to the cycles to failure of the material via different analytical/semi-analytical/phenomenological models [1–7]. A very well-established model is the following one:

$$\left(\frac{E}{E_0}\right)_i = \left(\frac{E}{E_0}\right)_{i,\mathrm{CA}} = A\left(\frac{N}{N_\mathrm{f}}\right)^b_{i,\mathrm{CA}} \tag{2}$$

where the subscript CA refers to constant amplitude tests, N is actual loading cycle, N_f is the cycle to failure of the material and A and b are constants of the specific material.

Equation (2) allows the determination of the remaining life of the material once the degradation of the material is known and vice versa. It is interesting to note that the behaviour of the material during a constant amplitude fatigue test depends only on the life fraction [5]. Therefore, it can be assumed that during a generic test at a constant stress level, the decay of the elastic properties with respect to the life fraction is the same regardless of the type of test (constant amplitude (CA) or stepwise test (SW)). Considering, moreover, that the reduction of the stiffness of the material occurs prevalently in the first load cycles, the following relationship can be considered valid:

$$\frac{E}{E_0}\left(\frac{N}{N_{bc}^{\mathrm{SW}}}\right)=\frac{E}{E_0}\left(\frac{N}{N_{\mathrm{f}}^{\mathrm{CA}}}\right) \tag{3}$$

Equation (3) indicates that at constant load, independent on the kind of test procedure (CA or SW), the Young's modulus loss is the same and depends on the cycles performed. In particular, N_{bc}^{SW} are the cycles of each loading block of a SW test (usually 10,000 cycles), while $N_{\mathrm{f}}^{\mathrm{CA}}$ are the cycles to failure of a CA test.

Equation (2) can also be used to estimate the stiffness degradation of the material during each single load block of the rapid fatigue test.

To estimate $\frac{E}{E_0}$, data from an extensometer or those deriving from the digital imaging correlation (DIC) can be used. Naturally, the first method is useful for laboratory applications, while the DIC technique requires an ad hoc equipment and setup and the presence of a surface applied speckle pattern useful for detecting displacements. An indirect method for estimating the loss of material stiffness is the one proposed by some researchers [9, 18], based on the study of the small temperature variations induced by the thermoelastic effect [19].

The phenomena affecting material mechanical behaviour are related to the heat energy involved in fatigue process; hence, mechanical data can be related to thermal data by considering that the formation of a new off-axis crack is a problem involving both variations of elastic energy and dissipations [4, 15, 19]. Indeed, the occurrence of cracks leads to a reduction of the structural stiffness which is related to the thermoelastic signal [9, 12, 18]. Furthermore, the energy released during damage (proportional to the energy needed to produce new fracture surfaces) [20] determines a change of internal energy and of work done on the material. Micro-cracks, however, are potential sites for the development of delamination between layers [20], and friction due to rubbing of damaged/delaminated surfaces also leads to energy dissipation [21].

All these phenomena influence the behaviour of the material by determining a variation of the thermoelastic signal and an increase in the dissipated energy; therefore, approaches based on the detection of the dissipated heat through measurement of the temperature signal can be useful for studying the damage by detecting these phenomena [7–17].

As shown in the recent work [22], during cyclic loading, a certain amount of energy supplied to the material is dissipated. In accordance with the first law of thermodynamics, the rate of mechanical energy supplied to an isolated system can be stored internally in the material or converted into heat [23, 24]. The latter contribution produces the temperature increase of the material or component [10–15].

Based on the total energy balance in terms of power per unit volume:

$$\dot{W}+\dot{Q}=\dot{U}=\dot{E}_{\mathrm{rev}}+\dot{E}_{\mathrm{irrev}}=\dot{E}_{\mathrm{s}}+\dot{E}_{\mathrm{d}}+\dot{E}_{\mathrm{rev}} \tag{4}$$

where $\dot{W}$ is the rate of mechanical energy supplied, $\dot{Q}$ is the rate of heat exchanged and $\dot{U}$ is the rate of internal energy variation. The rate of change of internal energy is composed of both a reversible part ($\dot{E}_{\mathrm{rev}}$) and an irreversible part ($\dot{E}_{\mathrm{irrev}}$). In particular, $\dot{E}_{\mathrm{rev}}$ represents the term due to the thermoelastic source which is related to the volume variations of the material induced by the imposed load, while $\dot{E}_{\mathrm{irrev}}$ can be divided into the contribution from intrinsic dissipations ($\dot{E}_{\mathrm{d}}$) and stored energy portion which is not converted into heat ($\dot{E}_{\mathrm{s}}$).

Focusing on the thermoelastic component, the energy can be determined with the following formula:

$$\dot{E}_{\mathrm{rev}}=f\,E_{\mathrm{rev}} \tag{5}$$

where f is the mechanical frequency of the tests. Instantaneous value of energy is:

$$E_{\mathrm{rev}}(t)=\rho c_p T_{\mathrm{the}}\sin(\omega t+\varphi) \tag{6}$$

where T_{the} is the amplitude of the thermoelastic signal and φ is the phase between the temperature and the reference load [9, 10] while ρ and c_p are, respectively, the density and the specific heat at constant pressure and ω is the pulsation which depends on the mechanical frequency. As presented in several papers [10–12], the monitoring of the thermoelastic signal can be correlated to the stiffness degradation of the material and can be used as a parameter to study the damage induced by fatigue processes and to estimate the failure cycles of the material. However, the thermoelastic signal is also influenced by the damage processes linked to the material stiffness loss (reduction of the normalized longitudinal elastic modulus) [9].

Considering that damage in composite materials begins once the load is applied so from the very first cycles and continues throughout the life of the material [25, 26], it is clear that a signal will be generated immediately, regardless of the type of load. This consideration is especially valid for the outer layers that are those monitored by the infrared camera detector.

For highly imposed stress values, the increase in the load leads to an increase in the thermoelastic signal as well. In this case, the effect of stress is more important than the reduction of stiffness. This is certainly true if the material presents a plateau in the stiffness degradation curve [12] as reported in [22]. Naturally, under these conditions, a dissipative thermal component is added to the thermoelastic signal in the first amplitude harmonic component [22]. Therefore, the amplitude of the first harmonic of the thermal signal, therefore, includes both effects related to the thermoelastic effect (also induced by dissipative processes) and effects deriving from irreversible processes. Moreover, since during a rapid fatigue test, the thermal signal does not depend only on cycles but also on the applied stress; under these conditions, the term representing the first harmonic component can be expressed as a function of the following sub-components:

$$T_1 = f\left(T^r_{\frac{E}{E0}}, T^r_\sigma, T_{1d}\right) \tag{7}$$

where $T^r_{\frac{E}{E0}}$ refers to the thermoelastic signal induced by the elastic properties loss, T^r_σ is the thermoelastic component linked to the increase in stress and T_{1d} is linked to fatigue dissipative processes.

Due to the variety of fatigue mechanisms that occur in the material and their random appearance affecting certain regions locally, one advantage of using thermal signal analysis is to locally 'inspect' the material for an effective and efficient monitoring of the damage.

Materials, Methods and Data Processing

The samples were made of carbon fibre-reinforced epoxy type resin and exhibited a nearly isotropic lamination sequence. Samples were made by the automated fibre placement (AFP) process. They were extracted from a $560 \times 695 \times 2.93$ mm^3 plate composed by 16 layers. The rectangular cross-section specimens have the following dimensions: $250 \times 25 \times 2.93$ mm^3 [22, 27]. The test campaign included static tensile tests, constant amplitude tests to construct the S/N curve and rapid fatigue tests (stepwise tests involving incremental loading blocks until material failure). The tests were performed on an INSTRON 8850 servo-hydraulic loading machine with a capacity of 250 kN.

The tensile tests, performed on five specimens (traverse speed of 1 mm/min), allowed the evaluation of the mechanical properties of the material: UTS 825 [MPa] (standard deviation 85 Mpa) and longitudinal Young's modulus 66 GPa (standard deviation 0.3 GPa). Constant amplitude fatigue tests (runout set at $2 * 10^6$ cycles) were performed at $R = 0.1$ and load frequency of 7 Hz. The detail of the CA tests is reported in the reference work [12].

The rapid fatigue tests were performed on three specimens under load control according to the procedure indicated in [10] at the same load ratio and mechanical frequency of the CA tests. The imposed loads ranged from 30%UTS to 85%UTS with a variable increment between one loading block and another.

During each loading block (whose duration was 10,000 cycles), three thermal sequences were acquired for each load block, i.e. in the first load block, thermal acquisitions were carried out at 3000/6000/9000 cycles. Thermal sequences were acquired by a FLIR X6540 SC cooled In-Sb sensor camera (640×512 pixel matrix array, NETD thermal sensitivity <30 mK) positioned in front of the loading machine (Fig. 1). The camera was set up to acquire sequences at 177 Hz. The test setup was set to obtain a spatial resolution of 0.35 mm/pixel. The equipment also included an extensometer with a measuring base of 25 mm for measuring the small deformations. The extensometer was connected to the data acquisition system integrated in the loading machine, with data acquired at a frequency of 100 Hz. The thermal signal acquisitions were simultaneous with those performed by the data acquisition system integrated in the loading machine.

In order to obtain the mechanical data (in terms of Young's modulus loss), the procedure for analysing data from the extensometer was similar to the one presented in [12]. Young's modulus was calculated as the average of the load/unload phase of a hysteresis cycle. Finally, that value, as described in [12], was referred to the first value of the data set for each stress level. The procedure for evaluating Young's modulus was identical for data obtained from classical fatigue tests (CA) and from tests at incremental loads (SW).

As for thermal signal, it was analysed using IRTA software that it is based on thermal signal reconstruction via least square method [10, 16]. The reference thermal model is:

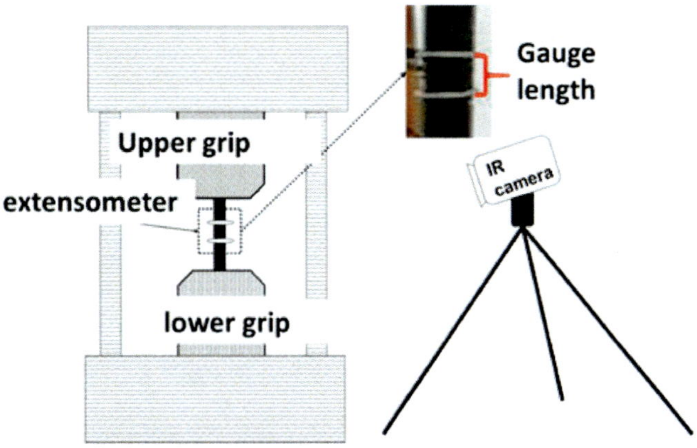

Fig. 1 Schematic of the setup and equipment

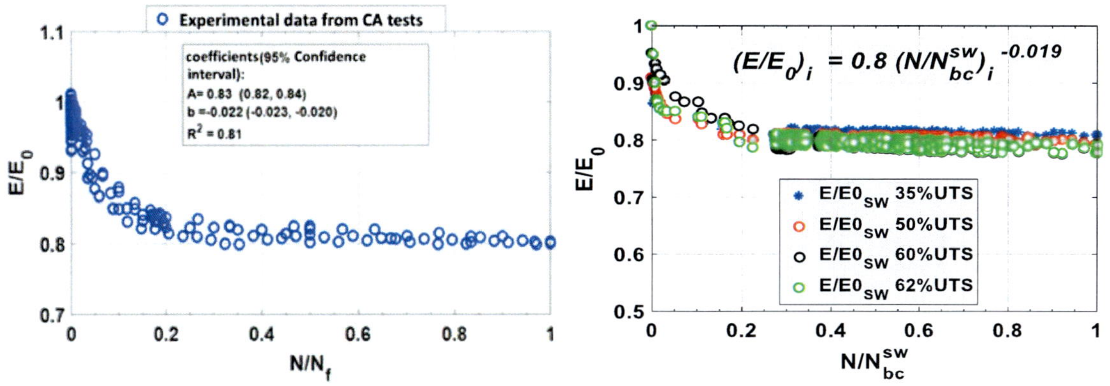

(a) (b)

Fig. 2 (**a**) $\left(\frac{E}{E_0}\right)_{i,\mathrm{CA}}$ and (**b**) $\left(\frac{E}{E_0}\right)_{i,\mathrm{SW}}$ curves versus cycle fractions [22]

$$T(t) = T_0 + at + T_1 \sin(\omega t + \varphi) + T_2 \sin(\omega t + \varphi_2) \tag{8}$$

where $T_0 + at$ represents the linear part of the model that influences the mean temperature increase while T_1 and T_2 are, respectively, the first and second amplitude harmonics (FAH, SAH) of thermal signal. The present work is focused only on FAH. The FAH acquired three times per loading block is under the form of 2D matrix of pixel, where each pixel represents at a specific location the value of the first harmonic component. All the maps were addressed to a spatial smoothing to reduce the noise in the measurement of FAH.

As reported in the work of Emery [9], and as found in different researches [12, 22], the minimum value in a map of FAH (usually referred to the gage length of the specimen) is directly related to stiffness loss. So, in the present research, the minimum values of T_1 were extracted from each map and normalized by the first value of the T_1 data series through the loading blocks. This ratio is represented by the parameter ζ_i [22].

Analysis

In this paragraph, the results in terms of mechanical and thermal data are reported. In Fig. 2a, it is possible to observe the evolution of the longitudinal Young's modulus E normalized by the reference value E_0. The data of all the tests were represented as a function of the fraction of life $\left(N/N_f^{\mathrm{CA}}\right)$. As can be seen, a single power law (Eq. 2) can represent all the data. The model fitting coefficients of the data (A, b) are reported in the same figure together with the 95% confidence interval.

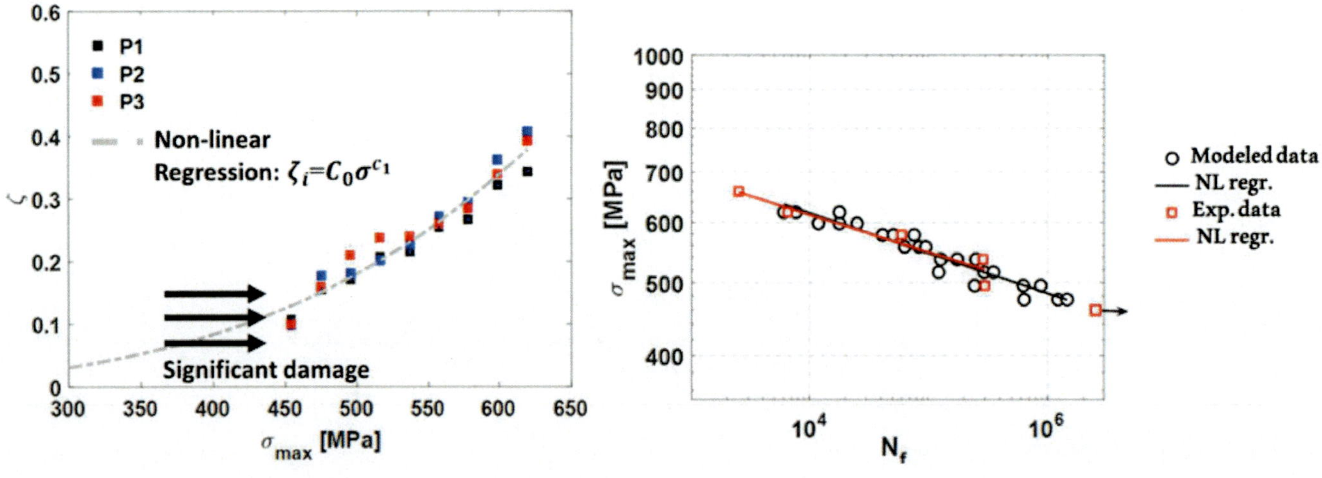

(a) (b)

Fig. 3 (**a**) Non-linear regression model of ζ_i data depending on the stress. (**b**) S/N curve obtained by model, compared to experimental data (double logarithmic axes)

Figure 2b instead represents the same quantities (*E/E₀*) relating to single load blocks of the test at incremental loads. The data in this case have been represented as a function of the fraction of life spent at a given load block ($N_{bc}^{SW} = 10^4$ cycles). As can be seen from Fig. 2b, the curves of the stiffness reduction for tested specimens at different stress levels show a similar trend which can be represented by a power law with the same fitting parameters of Eq.(2) and represented in Fig. 2a.

Referring to thermal data, it is possible to approximate the relationship between the thermal parameter based by FAH of thermal signal (ζ_i) and the stress (Fig. 3) with a power function whose coefficients and relative 95% confidence intervals are $C_0 = 8.439\text{e}^{-11}$ $(-1.189\text{e}^{-10}, 2.877\text{e}^{-10})$ and $c_1 = 3.457$ (3.077, 3.837). The coefficient of determination R^2 is 0.94. In Fig. 3a, the thermal considered data are related to higher stress levels where the damage is significant.

Since the relationship between $E/E_0 - \sigma$ is of the same type of the one that approximates $\zeta_i - \sigma$, one can assume that the relationship between $E/E_0 - \zeta_i$ is the same for all the tested samples, so it can be represented by the following model:

$$\left(\frac{E}{E_0}\right)_{i,\mathrm{SW}} = C_4(\zeta_i)^{c_5} \tag{9}$$

where C_4, c_5 are the constants depending on material.

Considering the validity of (3), as confirmed by Fig. 2a, b, using Eq. (9) to obtain the $\left(\frac{E}{E_0}\right)_{i,\mathrm{SW}}$, accounting for Eq. (2), it is easy to obtain the cycles to failure of a CA test from the following formula:

$$N_f^{\mathrm{CA}} = N/\left(\left(\frac{C_4}{A}\zeta_i^{c_5}\right)^{1/b}\right) \tag{10}$$

The comparison between the experimental data (red square markers) and data modeled with the presented procedure (black circular markers) is shown in Fig. 3b. In the Figure are also reported the data regression 'lines' and coefficients of the fitting. As can be seen from Fig. 3b, the adopted model approximates well the data deriving from calibration with a thermal parameter. In particular, the coefficients are almost the same, and furthermore the acquisition of thermal features from rapid tests allows for much more data to be available compared to a classic test, and this is advantageous both for the consistency of the results and for the definition of the S/N curve with more points.

Conclusion

In the present research, a new rapid procedure to evaluate the failure cycles of a quasi-isotropic CFRP was presented, based on FAH data and stiffness loss (longitudinal modulus) measurements. In particular, the thermal parameter used was the amplitude of the first harmonic of the thermal signal normalized with respect to the value of the first load block (ζ).

Finally, in this research:

- A new procedure has been presented to derive the S/N curve quickly.
- A damage parameter has been proposed which reflects the decay of the elastic properties of the material regardless of the type of test (classical or rapid test).
- Thanks to the calibration between thermal and mechanical data, it would also be possible to obtain maps of the 'stiffness degradation' of the material, which shows the decay of the material locally.

The disadvantage is that the procedure is currently valid for specimens and tests performed under laboratory conditions. As further development, the proposed procedure will be applied on a component during operating conditions in order to carry out structural health monitoring.

Acknowledgements This work is part of a R&D project 'SISTER CHECK -Sistema Termografico prototipale per il controllo di processo, la verifica e la caratterizzazione di materiali avanzati per l'aerospazio' of the research programme 'Horizon 2020' PON I&C 2014-2020 call.

References

1. Degrieck, J., Van Paepegem, W.: Fatigue damage modeling of fibre-reinforced composite materials: review. Appl. Mech. Rev. **54**(4), 279–299 (2001)
2. Carraro, P.A., et al.: Prediction of the crack density evolution in multidirectional laminates under fatigue loadings. Compos. Sci. Technol. **145**, 24–39 (2017)
3. Pakdel, H., Mohammadi, B.: Stiffness degradation of composite laminates due to matrix cracking and induced delamination during tension-tension fatigue. Eng. Fract. Mech. **216** (2019). https://doi.org/10.1016/j.engfracmech.2019.106489
4. Talreja, R., Varna, J.: Modeling Damage, Fatigue and Failure of Composite Materials Woodhead Publishing Series in Composites Science and Engineering, vol. 65. Elsevier (2016). https://doi.org/10.1016/C2013-0-16521-X
5. Poursaip, A., Ashby, M.F., Beaumont, P.W.R.: The fatigue damage mechanics of a carbon fibre composite laminate: I-development of the model. Comp. Sci. Technol. **25**, 193–218 (1986)
6. Huang, J., Pastor, M.J., Garnier, C., Gonga, X.J.: A new model for fatigue life prediction based on infrared thermography and degradation process for CFRP composite laminates. Int. J. Fatigue. **120**, 87–95 (2019)
7. Toubal, L., Moussa, K., Lorrain, B.: Damage evolution and infrared thermography in woven composite laminates under fatigue loading. Int. J. Fatigue. **28**, 1867–1872 (2006)
8. Montesano, J., Bougherara, H., Fawaz, Z.: Application of infrared thermography for the characterization of damage in braided carbon fiber reinforced polymer matrix composites. Compos. Part B. **60**, 137–143 (2014)
9. Emery, T.R., Dulieu-Barton, J.M.: Thermoelastic Stress Analysis of damage mechanisms in composite materials. Compos. Part A. **41**, 1729–1742 (2010)
10. Palumbo, D., De Finis, R., Demelio, P.G., Galietti, U.: A new rapid thermographic method to assess the fatigue limit in GFRP composites. Compos. Part B. **103**, 60–67 (2016)
11. De Finis, R., Palumbo, D., Galietti, U.: An experimental procedure based on infrared thermography for the assessment of crack density in quasi-isotropic CFRP. Eng. Fract. Mech. **258**, 108108 (2021)
12. De Finis, R., Palumbo, D., Galietti, U.: Evaluation of damage in composites by using thermoelastic stress analysis: a promising technique to assess the stiffness degradation. Fatigue Fract. Eng. Mater. Struct. **43**(9), 2085–2100 (2020)
13. Palumbo, D., Ancona, F., De Finis, R., Galietti, U.: Experimental study of the crack growth in stainless steels using thermal methods. Procedia Eng. **109**, 338–345 (2015)
14. La Rosa, G., Risitano, A.: Thermographic methodology for the rapid determination of the fatigue limit of materials and mechanical components. Int. J. Fatigue. **22**(1), 65–73 (2000)
15. Enke, N.F., Sandor, B.I.: Cyclic plasticity analysis by differential infrared thermography. In: Proc. Sixth International Congress on Experimental Mechanics, pp. 830–835, Portland, Oregon (1988)
16. Krapez, J.K., Pacou, D., Gardette, G.: Lock-in thermography and fatigue limit of metals. In: Quantitative Infrared Thermography, QIRT, Reims, France
17. De Finis, R., Palumbo, D.: Estimation of the dissipative heat sources related to the total energy input of a CFRP composite by using the second amplitude harmonic of the thermal signal. Materials **13**(12), 1–18, 2820 (2020)
18. Pitarresi, G., Found, M.S., Patterson, E.A.: An investigation of the influence of macroscopic heterogeneity on the thermoelastic response of fibre reinforced plastics. Comp. Sci. Technol. **65**, 269–280 (2005)
19. Stanley, P.: Beginnings and early development of thermoelastic stress analysis. Strain. **44**(4), 285–287 (2008)

20. Liu, S., Nairn, J.-A.: The formation and propagation of matrix microcracks in cross-ply laminates during static loading. J. Reinf. Plast. Compos. **11**, 158 (1992)
21. Parvizi, A., Garrett, K.W., Bailey, J.E.: Constrained cracking in CFRP cross ply laminates. J. Mater. Sci. **16**, 1785–1795 (1981)
22. De Finis, R., Palumbo, D., Galietti, U.: A new procedure for fatigue life prediction of CFRP relying on the first amplitude harmonic of the temperature signal. Int. J. Fatigue. **168**, 107370 (2023)
23. Nourian-Avval, A., Khonsari, M.M.: Rapid prediction of fatigue life based on thermodynamic entropy generation. Int. J. Fatigue **145**, 1–10, 106105 (2021)
24. Meneghetti, G., Ricotta, M.: Estimating the intrinsic dissipation using the second harmonic of the temperature signal in tension-compression fatigue. Part I-part II. Fatigue Fract. Eng. Mater. Struct. **44**(8), 2168–2185 (2021)
25. Nijssen, R.P.L.: Fatigue Life Prediction and Strength Degradation of Wind Turbine Rotor Blade Composites, p. 93. Knowledge Centre Wind Turbine Materials and Constructions (KC-WMC), The Netherlands (2006)
26. Whitworth, H.A.: Evaluation of the residual strength degradation in composite laminates under fatigue loading. Compos. Struct. **48**(4), 261–264 (2000)
27. Standard test method for tension-tension fatigue of polymer matrix composite materials D3479M-96

Rapid Fatigue Characterization via Infrared Thermography of AM-CM Composites

Nithinkumar Manoharan, Pharindra Pathak, Suhasini Gururaja, Vipin Kumar, and Uday Vaidya

Abstract A direct correlation exists between the microstructure of a composite (defect distribution, residual stresses, fiber geometry, and orientation) and the global composite mechanical behavior. Manufacturing processes govern the composite microstructure characteristics; thus, rapid process–microstructure–performance relationships must be established for emergent materials and manufacturing processes. In this chapter, a preform was manufactured via a modified Additive Manufacturing followed by Compression Molding (AM-CM) process with carbon fiber (CF) filled acrylonitrile butadiene styrene (ABS). The AM-CM plates were prepared with varying process parameters yielding distinct composite microstructures with varying porosity and fiber orientations. Micro-computed tomography has been used to characterize the microstructure of these composites. Preliminary mechanical testing (static) was conducted to ascertain these composites' static strength and stiffness properties. The current chapter strives to correlate the microstructure of these composites with their fatigue performance using rapid infrared thermography (IRT). A typical "staircase" loading has been adopted for IRT-based fatigue testing via self-heating. The stabilized temperature versus applied maximum stress profile plots yields a bi-linear curve indicating a pseudo-fatigue limit of each composite configuration. This bi-linear curve, coupled with stiffness degradation plots, can map composite microstructure with its fatigue performance. The approach outlined will provide a basis for rapidly characterizing and inserting emergent materials and manufacturing processes in fatigue-critical applications.

Keywords AM-CM composites · Self-heating · Fatigue limit · Infrared thermography

Introduction

Oakridge National Laboratory (ORNL) has recently developed a high-rate additive manufacturing process, namely, Additive Manufacturing (AM)—Compression Molding (CM) process that combines the classical CM process, which reduces porosity and creates functional surfaces, with the benefit of controlled fiber alignment made feasible in AM-printed parts [1]. High fiber alignment, minimal porosity, and careful bead placement can be precisely controlled to offer superior mechanical performance. A preform created using AM can include various materials for additional architectural functionalities, including over-molding, selective stiffening, and inserting electrically/thermally conductive channels. All these advantages can be achieved with a very quick part production cycle time, i.e., 2 minutes per part (1 ft × 1 ft part). As these AM-CM products transition to applications, the structural integrity of these composite structures has come to the forefront. Rapid certification of these structures needs innovative ways of characterizing their fatigue life. Conventional fatigue characterization processes use a heuristic S–N curve approach coupled with stiffness or strength degradation models that have proved to be time-

N. Manoharan · P. Pathak · S. Gururaja (✉)
Department of Aerospace Engineering, Auburn University, Auburn, AL, USA
e-mail: nzm0074@auburn.edu; pzp0057@auburn.edu; szg0130@auburn.edu

V. Kumar
Manufacturing Demonstration Facility, Oakridge National Lab, Knoxville, TN, USA
e-mail: kumarvi@ornl.gov

U. Vaidya
Tickle College of Engineering, University of Tennessee, Knoxville, TN, USA

Manufacturing Demonstration Facility, Oakridge National Lab, Knoxville, TN, USA
e-mail: uvaidya@utk.edu

C. Franck et al. (eds.), *Challenges in Mechanics of Biological Systems and Materials, Thermomechanics and Infrared Imaging, Time Dependent Materials and Residual Stress, Volume 2*, Conference Proceedings of the Society for Experimental Mechanics Series,
https://doi.org/10.1007/978-3-031-50470-9_8

consuming and agnostic of the material microstructure and inherent damage mechanics [2]. Recently, IRT-based approach to fatigue life characterization has shown promise as an attractive alternative [3, 4]. IRT uses the temperatures generated due to self-heating in composites under fatigue loading as an indicator of the amount of disorder or damage in a composite material, thus providing a credible measure of damage index related to the fatigue limit of the composite [5]. In this chapter, a staircase loading has been implemented to assess the fatigue limit of composites rapidly prepared using the AM-CM process using varying process parameters [6].

Background

Materials

The material system comprises acrylonitrile butadiene styrene (ABS) and carbon fiber (CF)-filled ABS pellets as feedstock. All specimens had an average 20% weight fraction of CF and a high fiber alignment. Figure 1 depicts the AM-CM process setup at ORNL. The final panels from the AM-CM processing had a ~8 mm thickness. Previous studies indicate a 50% reduction in volumetric porosity with high fiber alignment (approximately 82%) in the deposition direction based on X-ray micro-computed tomography (xCT) imaging of AM-CM samples [1]. A representative xCT image has been included in Fig. 2.

Fig. 1 AM-CM process

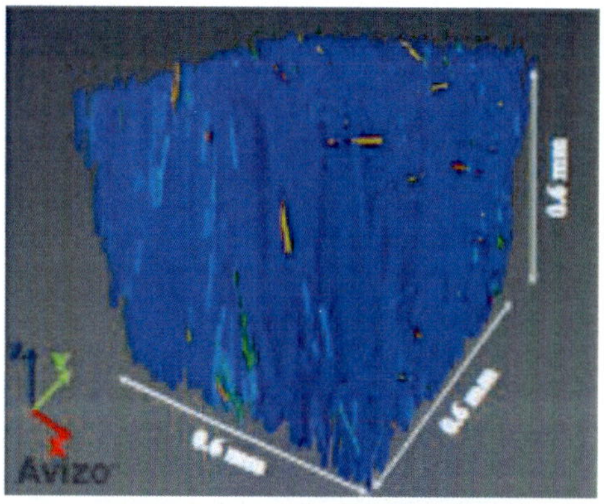

Fig. 2 A representative xCT image of the AM-CM sample [1]

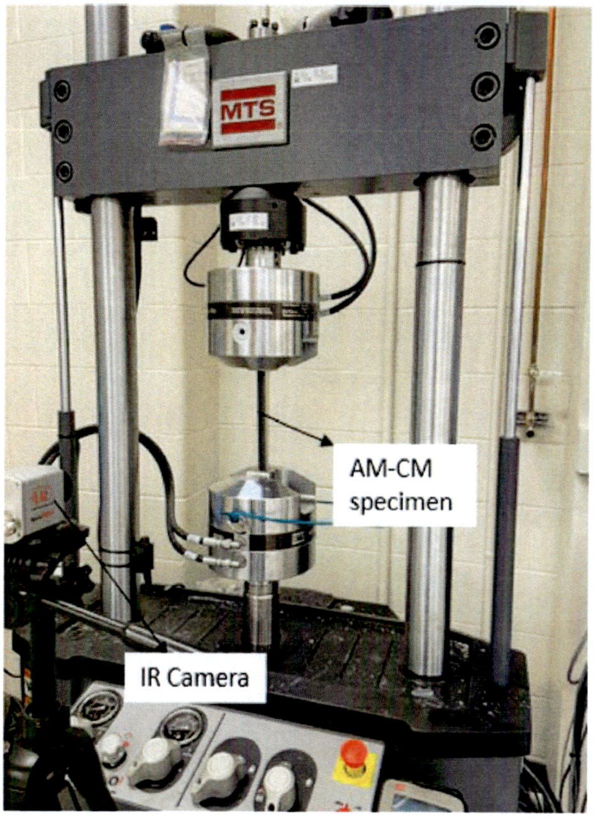

Fig. 3 Experimental setup for rapid fatigue characterization using IRT

Fatigue Limit Estimation Using IRT

All fatigue tests were carried out in a Landmark MTS servo-hydraulic system (Model 370.10) at a loading frequency of 15 Hz under tension–tension fatigue with a stress ratio R = 0.1 throughout testing (see Fig. 3). A high-performance IR camera (Microepsilon thermoIMAGER TIM QVGA) with an 18° lens, an optical resolution of 382 × 288 pixels, and 100 mK sensitivity was utilized in this chapter to record the surface temperature of the specimens throughout the tests. A staircase loading profile was used until all specimens failed, maintaining R = 0.1 at each step (see Fig. 4). The step size was taken as 4000 cycles based on temperature stabilization data from constant amplitude fatigue tests conducted on similar material systems.

As mentioned earlier, a rapid fatigue characterization technique using IRT was followed to characterize the specimens for fatigue [3, 6]. Static material data provided a baseline for choosing the appropriate staircase loading profile for fatigue tests. The fatigue limit is estimated by plotting the change in temperature with each load level against the maximum applied stress; it has been observed after a certain stress level, the temperature change is greater, and this threshold point could be used to estimate the fatigue limit of the material by obtaining a bi-linear fit. This greater temperature change could be related to an increased rate of damage accumulation that can be used as an indication of a fatigue limit.

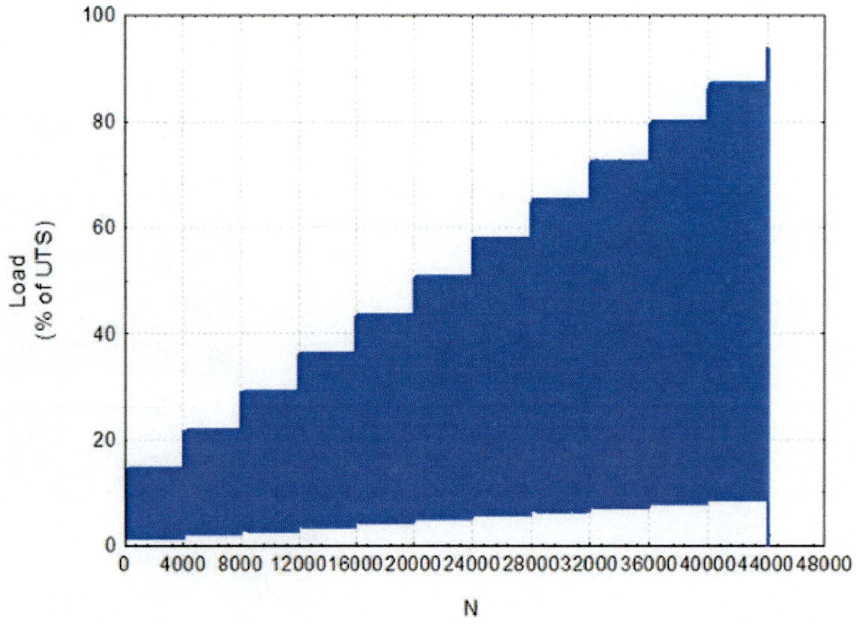

Fig. 4 Fatigue testing loading profile

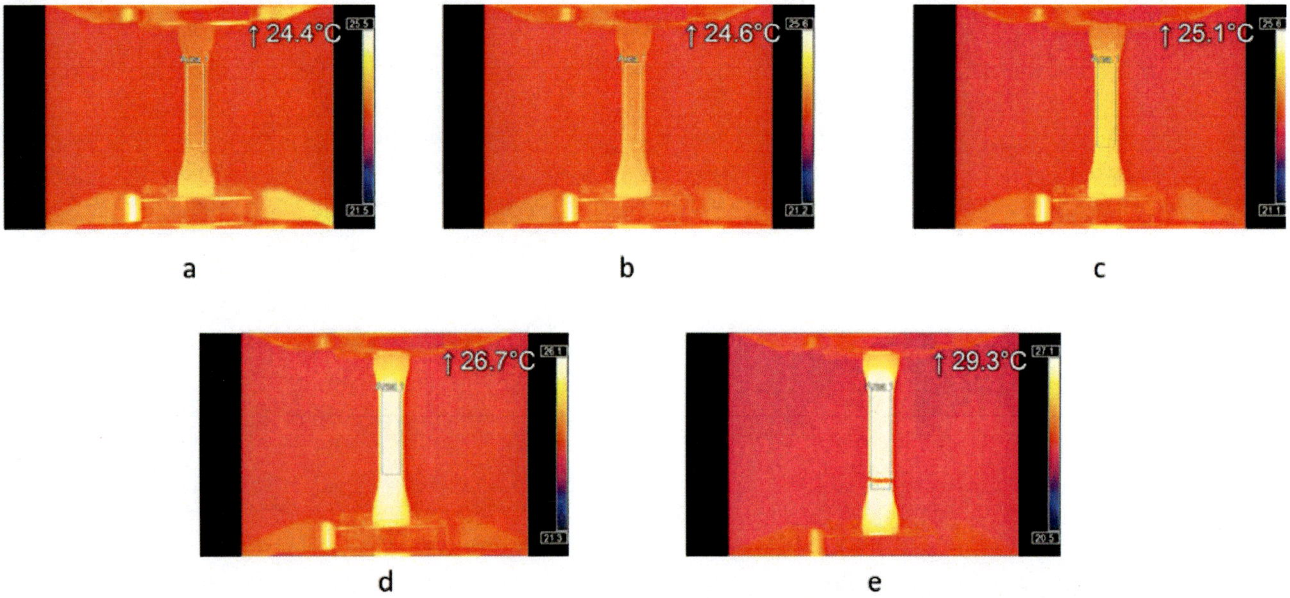

Fig. 5 Maximum temperature displayed within the region of interest for (**a**) 0 cycles, (**b**) 10,000 cycles, (**c**) 20,000 cycles, (**d**) 30,000 cycles, and (**e**) 39,750 cycles (failure)

Results and Discussion

Temperature evolution with respect to the number of cycles was recorded for each test until failure; the temperature contours with the respective number of cycles can be seen in Fig. 5. The number of cycles for each load level is maintained at 4000 cycles, ensuring temperature stabilization is attained. As the maximum stress amplitude increases with an increasing number of cycles, stiffness degradation is found to occur, which is obtained from the machine data (see Fig. 6). Stiffness can be used as a damage index quantifying damage evolution within the specimen throughout testing.

Figure 7 depicts the typical failure mode of specimens tested using the IRT-based fatigue approach. Further scanning electron microscopy (SEM) imaging of the failed surface is currently being carried out. From the temperature data obtained during the entire testing duration, the fatigue limit has been obtained by fitting a bi-linear curve delineating the two zones: zone 1 where the change in cyclic stabilized temperatures is well within 0.5 K and zone 2 exhibiting a higher cyclic stabilized

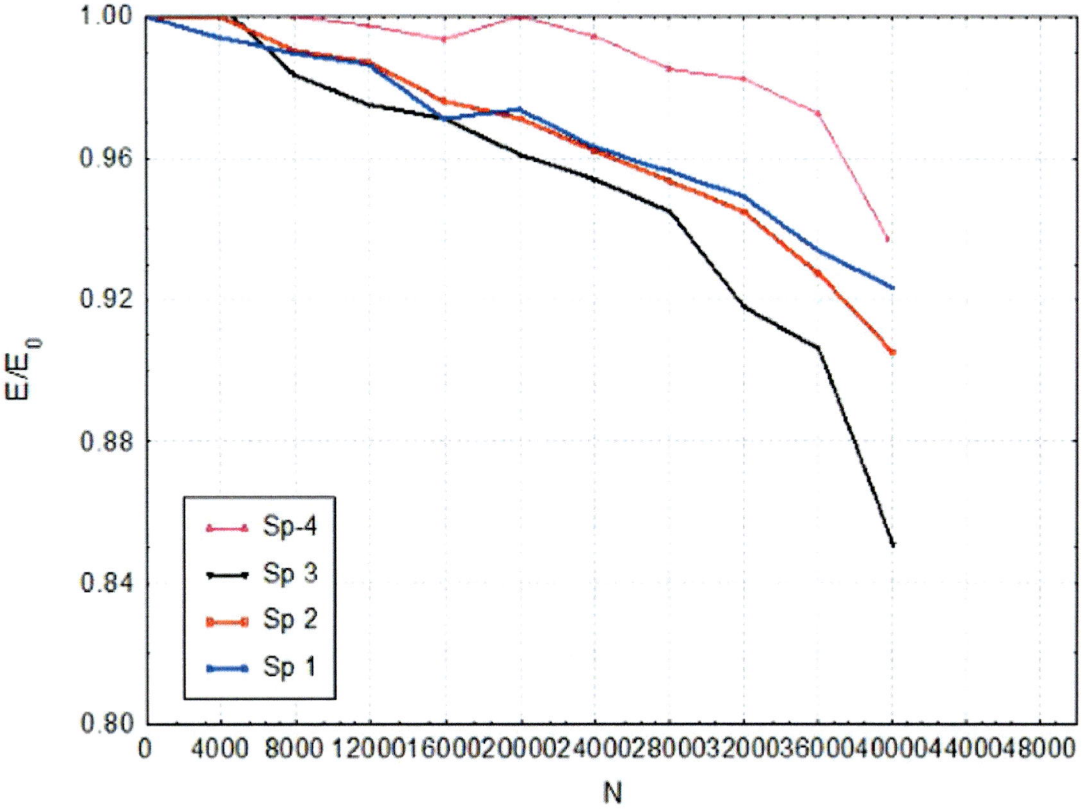

Fig. 6 Normalized stiffness degradation versus the number of cycles for AM-CM samples testing via IRT rapid fatigue technique

Fig. 7 Failed specimens tested using IRT fatigue

temperature. The transition region between the two zones indicates a threshold stress level below which no significant damage occurs in the specimen. This threshold can be viewed as the fatigue limit of the material. Figure 8 depicts the stabilized cyclic temperatures for four specimens tested under the same load profile (Fig. 4). As can be seen in Fig. 8, a distinct change in the slopes of cyclic stabilized temperatures during fatigue appears between 60 and 70% σ_{UTS}. On average, the fatigue limit of these specimens can be specified as 66.52% σ_{UTS}.

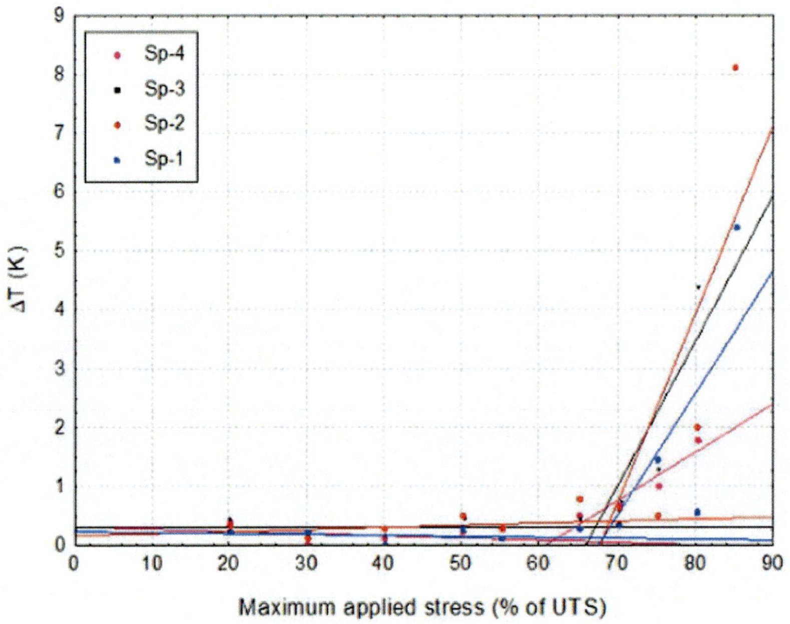

Fig. 8 Fatigue limit estimation using bi-linear fits of cyclic stabilized temperature versus $\sigma_{max}(\%of\,UTS)$

Conclusion

Fatigue characterization of novel CF/ABS composites processed using AM-CM technique was carried out using IRT. Repeatable temperature profiles were obtained indicating a physical phenomenon correlating fatigue damage accumulation and surface temperature profiles. The duration of fatigue testing has been greatly reduced validating our hypothesis of using temperature as a means of rapid fatigue life characterization. Efforts are underway to correlate process parameters of the AM-CM technique with inherent microstructural/morphological variations and resulting fatigue life of these unique composites via IRT.

References

1. Kumar, V., et al.: High-performance molded composites using additively manufactured preforms with controlled fiber and pore morphology. Addit. Manuf. **37**, 101733 (2021)
2. Vassilopoulos, A.P.: The history of fiber reinforced polymer composite laminate failure. Int. J. Fatigue **134**, 105512 (2020)
3. Montesano, J., Fawaz, Z., Bougherara, H.: Use of infrared thermography to investigate the fatigue behavior of a carbon fiber reinforced polymer composite. Compos. Struct. **97**, 76–83 (2013)
4. Huang, J., et al.: Rapid evaluation of fatigue limit on thermographic data analysis. Int. J. Fatigue **104**, 293–301 (2017)
5. Mohammadi, B., et al.: Damage-entropy model for fatigue life evaluation of off-axis unidirectional composites. Compos. Struct. **270**, 114100 (2021)
6. Fargione, G., et al.: Rapid determination of the fatigue curve by the thermographic method. Int. J. Fatigue **24**, 11–19 (2002)

Condition Assessment by Thermal Emission (CATE) for In Situ Monitoring of Fatigue Crack Growth

Khurram Amjad, Peter R. Lambert, Ceri A. Middleton, Richard J. Greene, and Eann A. Patterson

Abstract The cost and size of instrumentation for thermoelastic stress analysis (TSA) have often been an inhibiting factor for its use in industrial applications. This has been alleviated to some extent by the development of packaged infrared (IR) bolometers which have become popular for non-destructive evaluation of structures. Recent work has demonstrated that an original equipment manufacturer (OEM) microbolometer, combined with a single circuit board with dimensions equivalent to a credit card, can be used to detect cracks of the order of 1 mm long and to monitor their propagation. The monitoring system costs about one tenth the price of a packaged bolometer and can provide results in quasi real time without the need for calibration. The system uses the principles of TSA to acquire thermal images and evaluate the amplitude of the thermal signal over the field of view, i.e., a map of thermal emission amplitude. Feature vectors are extracted from the time-varying maps of thermal emission amplitude and used to identify changes in them that occur when a crack initiates or propagates in the field of view. Since the technique does not generate TSA data but uses uncalibrated thermal emission data, it has been named Condition Assessment by Thermal Emission (CATE). The CATE system's crack detection capability has been evaluated in laboratory conditions and compared against a state-of-the-art IR photovoltaic effect detector. It was demonstrated that the CATE system is capable of detecting cracks as small as 1 mm at loading frequencies as low as 0.3 Hz. Evaluations in industrial conditions on large-scale structures are being concluded and imply that there will be little loss of capability in the more demanding applications.

Keywords Condition monitoring · Damage detection · Thermoelastic stress analysis · Fatigue · Microbolometer

Introduction

Non-destructive evaluation (NDE) or non-destructive testing (NDT) methods allow monitoring of engineering structures without the technique itself causing changes to that structure. There are a large range of both point and full-field NDE methods which can be applied in both laboratory and industrial settings, depending on the conditions, materials, and desired information collected [1–3]. Thermoelastic stress analysis (TSA) is one full-field technique that uses the thermoelastic effect, combining infrared imagery and information on the applied load, to produce surface maps which are proportional to the sum of the principal stresses. TSA has previously been used for damage detection in metals [4–6] and composites [7–9]. Until recently, the high cost, volume, and mass of infrared sensors have inhibited use in industrial environments; however, developments in infrared camera technology have resulted in cheaper, lower volume packaged microbolometer systems which are more suited to use in industrial settings [10]. A further decrease in cost and size has occurred with the development of cm size OEM (original equipment manufacturer) board cameras, such as the Lepton Micro Thermal Camera Module (FLIR, Wilsonville, OR, USA), which cost on the order of $100s, rather than the $1000s to $10,000s of previous technology. The decrease in cost and size results in an associated decrease in sensitivity and resolution; however by combining the data

K. Amjad
UK Atomic Energy Authority, Oxfordshire, UK

P. R. Lambert · C. A. Middleton · E. A. Patterson
School of Engineering, University of Liverpool, The Quadrangle, Brownlow Hill, Liverpool, UK
e-mail: eann.patterson@liverpool.ac.uk

R. J. Greene
Strain Solutions Ltd, Dunston Innovation Centre, Chesterfield, Derbyshire, UK

C. Franck et al. (eds.), *Challenges in Mechanics of Biological Systems and Materials, Thermomechanics and Infrared Imaging, Time Dependent Materials and Residual Stress, Volume 2*, Conference Proceedings of the Society for Experimental Mechanics Series,
https://doi.org/10.1007/978-3-031-50470-9_9

from these sensors with data processing techniques, these sensors can be used for damage detection in aerospace materials [11, 12].

Methods

A simple hole-in-plate specimen was manufactured from an aerospace Al alloy, with dimensions 200 mm length, 40 mm width, 1.6 mm thickness, and a central hole of 6 mm diameter. The central portion of the front surface of the specimen was painted with aerospace primer paint (LAS780-001, LAS Aerospace Ltd, UK), and a resistance strain gauge was installed on an area out of the camera's field of view. Specimens were fatigued using cyclic loading between 0.875 and 8.75 kN at 10 Hz for 80,000 cycles to accelerate crack initiation before data collection. Specimens were then loaded with either sinusoidal cyclic loading at 1.2 Hz or an idealized flight cycle loading spectrum. Each flight cycle loading was followed by five cycles of sinusoidal loading at 1.5 Hz, with the sequence then repeated until failure.

Two camera systems were used for data collection during constant sinusoidal loading, a high-resolution photovoltaic detector (FLIR SC7650, Teledyne FLIR LLC, USA) with DeltaTherm software (Stress Photonics, USA) producing TSA maps and a bespoke OEM microbolometer system (Fig. 1). During flight cycle loading, data was collected using the OEM microbolometer system only. The OEM microbolometer system combines a Lepton board camera with a Raspberry Pi control computer and requires input from a strain gauge on the structure of interest. Bespoke software converts the strain information and infrared arrays to maps of the thermoelastic response, called Condition Assessment by Thermal Emission (CATE) data, to distinguish them from the quantitative TSA data generated by traditional approaches [12].

TSA and CATE maps were generated by the two systems in real time, and CATE data were also processed in near real time on the Raspberry Pi control computer using image decomposition [13, 14]. The two-dimensional images were decomposed to generate a one-dimensional feature vector, and then the Euclidean distances between the feature vectors representing the initial undamaged state and the maps over time were calculated [12]. The same image decomposition approach was applied to the TSA maps from the photovoltaic detector in post-processing. For flight cycle data, the algorithm extracts the response for different sections of the flight cycle, and so a Euclidean distance between the feature vector for each component of the applied load spectrum is calculated.

Fig. 1 Experimental set-up for thermal data collection. Photovoltaic detector (left) and CATE system (center, foreground), specimen with aerospace primer (right). (Adapted from [12])

Results and Discussion

The normalized values of Euclidean distance between feature vectors representing data from initial and current states from both the photovoltaic detector and the OEM system during constant amplitude loading are shown in Fig. 2, along with inset figures of the TSA and CATE maps. Due to the low-frequency loading, the TSA and CATE maps do not show clear hotspots associated with crack tips, which would be present under adiabatic conditions [5, 15–17]. However, there are changes in the overall data distribution that can be quantified using the image decomposition approach, which are indicated by increases in the Euclidean distance. An initial plateau at approximately zero is seen in both graphs, with a slight increase visible in the photovoltaic detector data over time. A significant increase in the Euclidean distance is seen at ~7.5 hours in both detectors.

Crack detection is defined to occur when the Euclidean distance between feature vectors representing the initial and current state increases significantly above a threshold that is calculated as a function of the rolling median of the Euclidean distance [12]. This crack detection instant is indicated earlier in the data from the photovoltaic detector than the OEM detector (Fig. 2), corresponding to sub-mm crack lengths. As the resolution of the photovoltaic detector is higher than the microbolometer, this earlier detection is not surprising. In real terms, the OEM detector is indicating damage at the beginning of crack propagation, at a crack length of ~1–2 mm, which is an order of magnitude shorter than cracks commonly located by visual inspection in the aerospace industry. Therefore, in comparison with a photovoltaic detector system for use in an industrial setting, the slight delay in crack detection would likely be outweighed by the significant decrease in cost (from $10,000s to $100s) and the smaller footprint of the system.

CATE maps were also collected during flight cycle loading (Fig. 3), with particular areas in the loading cycle automatically identified by the software. The image decomposition method was applied to results from these individual sections, as well as a "broadband" effect from the combined waveforms. Certain regions of the flight cycle showed no significant changes in the Euclidean distance (Fig. 3, center-right). However, other components show similar increases over time to those seen with the sinusoidal loading, indicative of changes in the damage state (Fig. 3, top and bottom-right). Importantly, the "broadband" signal also shows the increase in Euclidean distance, demonstrating that this approach would also work for a random loading situation. Analysis of the individual Euclidean distance plots related to individual components of the flight cycle indicates that these changes can be identified at frequencies as low as 0.3 Hz at high enough strain amplitudes [12].

The prototype system has also been installed in full-scale aerospace tests at industrial sites. The successful collection of data from these installations shows a proof of concept of installation and use by personnel in industrial laboratory environments. Future work will build on these activities for damage detection and monitoring in these environments.

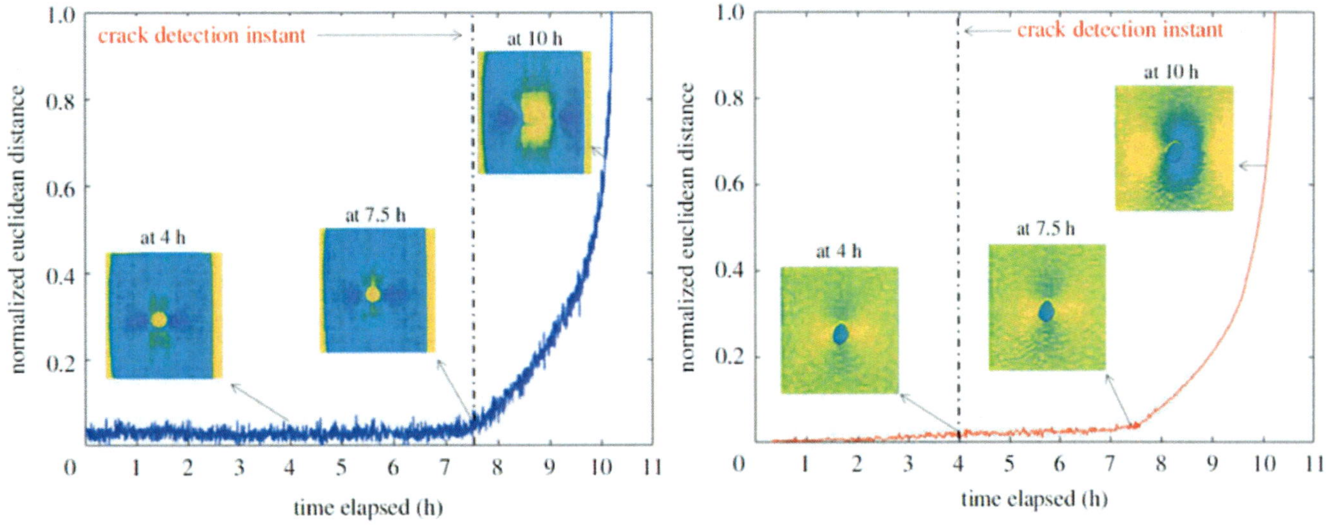

Fig. 2 Normalized Euclidean distance between feature vectors representing the initial and current state from the CATE system, processed in quasi real time (left) and post-processed Euclidean distance from the photovoltaic detector (right). Insets show CATE and TSA maps at given times. (Adapted from [12])

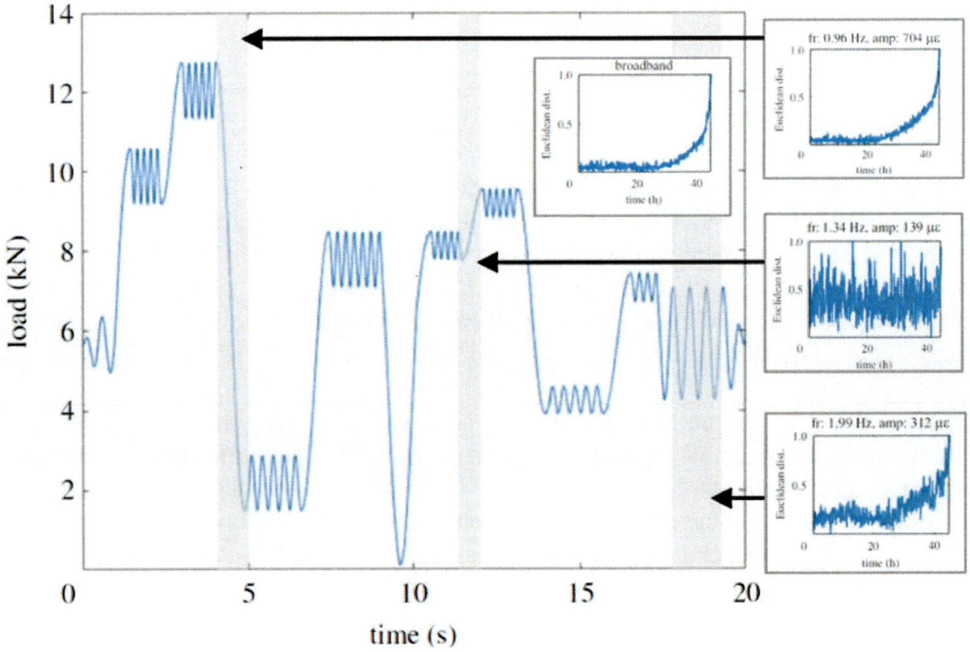

Fig. 3 Applied flight cycle load, calculated normalized Euclidean distances between feature vectors representing the initial and current state (right) for selected components of the spectrum (indicated by gray shading), and broadband thermoelastic response (inset). (Adapted from [12])

Conclusion

A low-cost, low-volume thermal emission-based system has been developed for use in damage detection. The system collects infrared data, processing it in near real time to CATE (Condition Assessment by Thermal Emission) maps using an on-board credit card-sized computer. CATE maps are then further processed to give a quantitative measure of the damage through image decomposition techniques. Detection of fatigue cracks is possible at 1–2 mm length under laboratory conditions and at loading frequencies as low as 0.3 Hz. The system has been successfully installed in full scale aerospace tests in industrial environments.

Acknowledgments This study was part of the DIMES (Development of Integrated Measurement Systems) project which has received funding from the Clean Sky 2 Joint Undertaking under the European Union's Horizon 2020 research and innovation program under grant agreement no. 820951. The opinions expressed in this article reflect only the authors' view, and the Clean Sky 2 Joint Undertaking is not responsible for any use that may be made of the information it contains. The authors acknowledge the productive discussions with their partners and topic manager in the DIMES project, including Erwin Hack of Empa and Linden Harris of Airbus.

References

1. Sakagami, T.: Remote nondestructive evaluation technique using infrared thermography for fatigue cracks in steel bridges. Fatigue Fract. Eng. Mater. Struct. **38**(7), 755–779 (2015). https://doi.org/10.1111/ffe.12302
2. Rastogi, P., Hack, E. (eds.): Optical Methods for Solid Mechanics. Wiley-VCH, Weinheim (2012)
3. Epstein, J.S.: Experimental Techniques in Fracture. VCH, New York (1993)
4. Patki, A.S., Patterson, E.A.: Thermoelastic stress analysis of fatigue cracks subject to overloads. Fatigue Fract. Eng. Mater. Struct. **33**(12), 809–821 (2010). https://doi.org/10.1111/j.1460-2695.2010.01471.x
5. Diaz, F.A., Patterson, E.A., Tomlinson, R.A., Yates, J.R.: Measuring stress intensity factors during fatigue crack growth using thermoelasticity. Fatigue Fract. Eng. Mater. Struct. **27**, 571–583 (2004). https://doi.org/10.1111/j.1460-2695.2004.00782.x
6. Palumbo, D., Galietti, U.: Characterisation of steel welded joints by infrared thermographic methods. Quant. InfraRed Thermogr. J. **11**, 29–42 (2014). https://doi.org/10.1080/17686733.2013.874220
7. Emery, T., Dulieu-Barton, J.: Thermoelastic stress analysis of damage mechanisms in composite materials. Compos. Part A Appl. Sci. Manuf. **41**, 1729–1742 (2010). https://doi.org/10.1016/j.compositesa.2009.08.015
8. Krstulovic-Opara, L., Klarin, B., Neves, P., Domazet, Z.: Thermal imaging and thermoelastic stress analysis of impact damage of composite materials. Eng. Fail. Anal. **18**(2), 713–719 (2011)

9. Shiozawa, D., Sakagami, T., Nakamura, Y., Nonaka, S., Hamada, K.: Fatigue damage evaluation of short carbon fiber reinforced plastics based on phase information of thermoelastic temperature change. Sensors. **17**, 2824 (2017). https://doi.org/10.3390/s17122824
10. Rajic, N., Street, N.: A performance comparison between cooled and uncooled infrared detectors for thermoelastic stress analysis. Quant. InfraRed Thermogr. J. **11**(2), 207–221 (2014). https://doi.org/10.1080/17686733.2014.962835
11. Middleton, C.A., Weihrauch, M., Christian, W.J.R., Greene, R.J., Patterson, E.A.: Detection and tracking of cracks based on thermoelastic stress analysis: detection and tracking of cracks. R. Soc. Open Sci. **7**(12) (2020). https://doi.org/10.1098/rsos.200823
12. Amjad, K., Lambert, P., Middleton, C.A., Greene, R.J., Patterson, E.A.: A thermal emissions-based real-time monitoring system for in situ detection of fatigue cracks. Proc. R. Soc. A Math. Phys. Eng. Sci. **478**(2266) (2022). https://doi.org/10.1098/rspa.2021.0796
13. Mukundan, R., Ong, S., Lee, P.: Image analysis by Tchebichef moments. IEEE Trans. Image Process. **10**(9), 1357–1364 (2001)
14. Yap, P., Paramesran, R., Ong, S.: Image analysis using Hahn moments. IEEE Trans. Pattern Anal. Mach. Intell. **29**(11), 2057–2062 (2007)
15. Rajic, N., Brooks, C.: Automated crack detection and crack growth rate measurement using thermoelasticity. Procedia Eng. **188**, 463–470 (2017). https://doi.org/10.1016/j.proeng.2017.04.509
16. Middleton, C.A., Gaio, A., Patterson, E.A., Greene, R.J.: Towards automated tracking of initiation and propagation of cracks in aluminium alloy coupons using thermoelastic stress analysis. J. Nondestruct. Eval. **123**(38), 18 (2019). https://doi.org/10.1007/s10921-018-0555-4
17. Thatcher, J., Crump, D., Devivier, C., Bailey, P., Dulieu-Barton, J.: Low cost infrared thermography for automated crack monitoring in fatigue testing. Opt. Lasers Eng. **126**, 105914 (2020). https://doi.org/10.1016/j.optlaseng.2019.105914

Effect of Geometrical Discontinuities on Mode Shapes Stress Maps Using Infrared Thermography

Ángel J. Molina-Viedma, Davide Palumbo, Rosa De Finis, Francesca Di Carolo, Luis Felipe-Sesé, Elías López-Alba, Umberto Galietti, and Francisco A. Díaz

Abstract Thermoelastic stress analysis (TSA) is an experimental technique that provides maps of stress variation in loaded solids related to the variation of temperature, under linear and adiabatic conditions. Traditionally, TSA has been employed in mechanical components under monotonic harmonic excitation measuring with infrared thermography. Hence, the amplitude of the temperature signal in a pixel can be related to the stress amplitude, after proper calibration of the thermoelastic properties. The main purpose so far has been the analysis of stress maps for fatigue and fracture mechanics by applying a load at a low frequency, about 10 Hz, but high enough to avoid heat conduction. However, recent studies have exploited harmonic excitation for a new use, which is the evaluation of stress maps associated with mode shapes. These studies have shown the capability to provide the shape information at much higher frequencies, of hundreds and even thousands of Hertz. In this work, the authors evaluate the capabilities of this methodology, based on harmonic excitation of mechanical components, to identify the effect of discontinuities on the stress maps associated with mode shapes. It is based on the fact that each mode shape is unique, changing the stress distribution from one to another. For this purpose, different kinds of geometrical discontinuities on flat components were evaluated. First, their natural frequencies were identified, and, afterwards, a harmonic excitation was applied at those frequencies. The response of the specimen was recorded with an infrared thermocamera. The analysis of the thermal signals, related to stress, reveals that how much the maps are altered depends on the location of the discontinuity regarding the modal shape of the sound specimen. In this way, with a set of some mode shapes, it is possible to identify, at least, one that highlights the effect of discontinuities.

Keywords Thermoelasticity · Mode shape · Stress concentration

Introduction

Thermoelastic stress analysis (TSA) is an experimental, contactless, full-field technique used to assess the superficial stresses on components subjected to dynamic loads [1, 2]. The thermoelastic effect can be described assuming linear elastic and adiabatic conditions and can be resumed by a direct relationship between the sum of the principal stresses (first stress invariant) and the temperature as a response of the material [3].

Thanks to its versatility, TSA has been used, in the last years, in many applications concerning the non-destructive evaluation of components [4], crack growth monitoring [5], and the validation of finite element models (FEM) [6]. Specific applications also regarded the assessment of the fatigue limit or, more in general, the fatigue behavior of metallic and composite materials [7].

All the applications described above have shown the potentiality of the TSA technique for in situ applications on real components subjected to actual loading conditions. In this regard, the in-lab tests were usually carried out by imposing a sinusoidal load with a suitable mechanical frequency that depends on the tested material, for assuring the adiabatic conditions [1–3]. However, the real loads that act on real components have a random behavior and could cover a wide

Á. J. Molina-Viedma (✉) · L. Felipe-Sesé · E. López-Alba · F. A. Díaz
Department of Mechanical and Mining Engineering, Universidad de Jaén, Jaén, Spain
e-mail: ajmolina@ujaen.es; lfelipe@ujaen.es; elalba@ujaen.es; fdiaz@ujaen.es

D. Palumbo · R. De Finis · F. Di Carolo · U. Galietti
Department of Mechanics, Mathematics and Management, Politecnico di Bari, Bari, Italy
e-mail: davide.palumbo@poliba.it; rosa.definis@poliba.it; francesca.dicarolo@poliba.it; umberto.galietti@poliba.it

C. Franck et al. (eds.), *Challenges in Mechanics of Biological Systems and Materials, Thermomechanics and Infrared Imaging, Time Dependent Materials and Residual Stress, Volume 2*, Conference Proceedings of the Society for Experimental Mechanics Series,
https://doi.org/10.1007/978-3-031-50470-9_10

range of mechanical frequencies. In the past, some authors proposed approaches capable of extracting the stress information in terms of superficial stress maps for components subjected to random loads [3, 8].

Considering that TSA with infrared thermography works under periodic excitation, it can be also employed for a different kind of studies like estimation of stress mode shapes obtained through sinusoidal excitation. The specimen is excited sinusoidally at one of its resonances, and the thermal response of the specimen can be registered with an IR camera. The amplitude of the thermal map corresponding to the frequency of excitation is proportionally related to the stress field associated to the mode shape of the excited resonance, as the frequency is high enough to meet adiabatic conditions. This is a new approach so that only a few works have explored this line. Preliminary works were performed [9, 10] in turbine blades or aerospace composite components [11]. Phan et al. [12] showed that thermal conductivity affects the measurement in metallic materials for low-frequency modes, below 20 Hz, where the correlation with the theoretical and numerical models failed. A more detailed parameter description was performed by Molina-Viedma et al. [13], focusing on high frequencies up to 2000 Hz. In this work, a finite element model was employed to validate the stress maps. The evolution of the thermal signal with frequency was studied, proving that the signal decreases less severely than displacements due to the higher stiffness of high-order modes. In a later work, Molina-Viedma et al. [14] proposed an alternative to Fourier transform based on the periodogram Welch's averaging method for multi-harmonic signals. This method allows to obtain true amplitude and phase shift of the thermoelastic signals. It is more computationally efficient and reduces the results' size for storage.

In this work, the thermoelastic effect was employed to analyze the influence of geometric discontinuities on stress maps due to its importance for structural integrity. An aluminum beam with a lateral notch was excited at their resonances, and thermoelastic signal was recorded with an IR camera. A finite element method (FEM) model was employed to identify the modes whose stress maps are more affected by the discontinuity.

Experiments

The experimental campaign was designed to demonstrate the applicability of the TSA technique in modal analysis and its ability to identify stress concentration areas. Additionally, the experimental data was compared with FEM models. All tests were carried out on a test bench equipped as shown in Fig. 1. The sample was clamped with a fixed constraint at the base, and the mechanical vibration was transmitted through this constraint using an electrodynamic shaker.

The experimental tests were carried out on a specimen made of Aluminum 2024 series. The dimensions were 50 mm wide and 5 mm thick. The length was designed so that the free length of the specimen after fixing it to the shaker was 420 mm. A rectangular side notch was made at the middle of its length, as shown in Fig. 2.

The first phase of experimental activity involved the identification of natural frequencies for the components. This activity was carried out by monitoring the applied excitation at the fixture with an accelerometer and the response of the beam at the tip with a laser Doppler vibrometer, following impulsive mechanical excitation. The natural frequencies are reported in Table 1. After identifying the natural frequencies, we proceeded to identify, for each natural frequency, the maximum excitation amplitude that could be applied. The mechanical excitation applied was a constant sinusoidal excitation, and the limit for this amplitude was established based on the limits imposed by the mechanical excitation hardware. The resulting amplitude at the fixture, d, is included in Table 1.

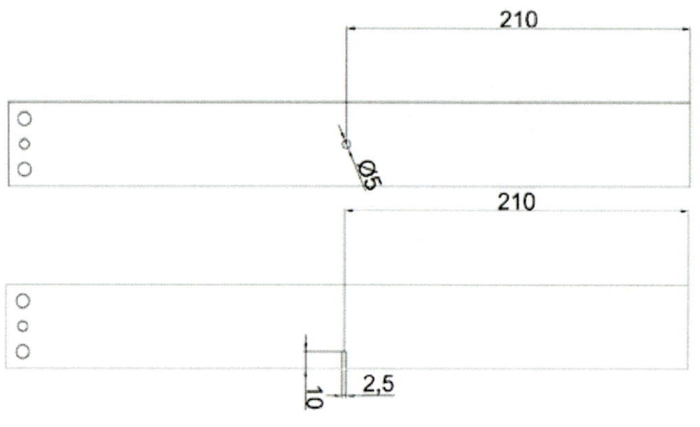

Fig. 1 Test bench scheme

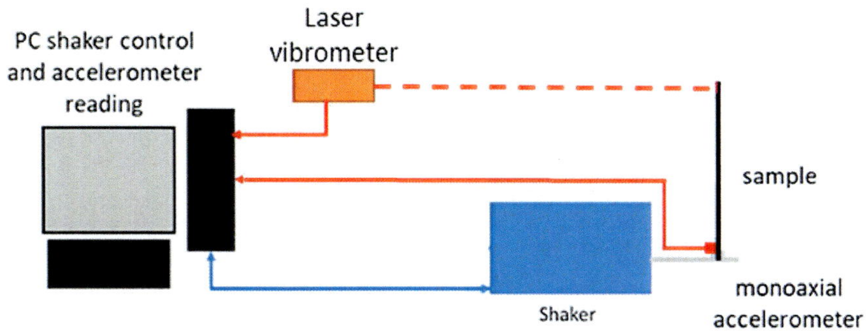

Fig. 2 Position and dimensions of the notch in the beam aluminum specimen

Table 1 Excitation frequency and IR camera acquisition parameters

Freq (Hz)	Samp rate (Hz)	Alias (Hz)	IT (μs)	IT (%)	d (mm)
125	121	4	2700	33.75	0.4
390	197	4	890	34.71	0.1
435	219	3	800	34.8	0.1
984	245	4	400	39.36	0.01
1476	296	4	270	39.852	0.08

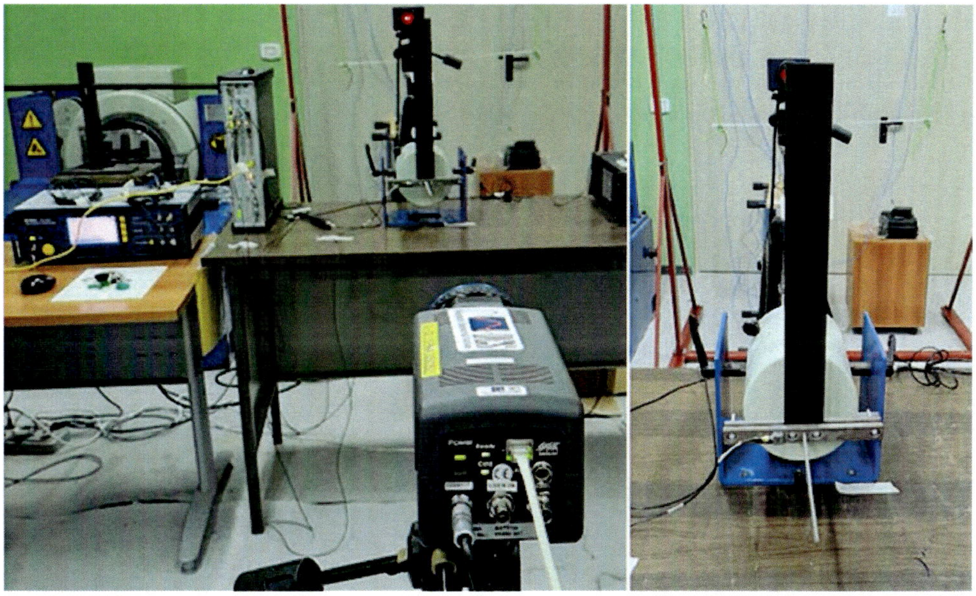

Fig. 3 Test set up showing the camera in front of the specimen, the laser behind it pointing at the tip and the accelerometer at the fixture

During this excitation, thermal sequences were acquired using a FLIR X6540sc cooled thermal camera, which ensured the ability to acquire at high frame rates (up to about 2000 Hz) and a higher sensitivity in temperature measurement (about 20 mK). The IR camera was equipped with two different lenses depending on the distance between the sensor and the part. Specifically, two different frames were realized: the first characterized by a mm/pixel ratio of 0.85, which required the use of a 50 mm lens and provided a complete view of the entire specimen, and the second frame characterized by a mm/pixel ratio of 0.16, which required the use of a 25 mm lens. The second frame acquisitions were carried out to improve the resolution of the thermal data around the imposed defects. The thermal acquisitions and data analysis were carried out using the IRTA2 software. The setup described is shown in Fig. 3.

The acquisition frequency was determined based on the undersampling strategy applied. This strategy involves reconstructing the sinusoidal signal through undersampling characterized by a frequency that deviates slightly from a submultiple of the frequency of the observed phenomenon. Thanks to the aliasing between the submultiple of the excitation frequency and the acquisition frequency, it is possible to reconstruct a sinusoidal signal characterized by the same amplitude as the observed signal and a frequency equal to the aliasing frequency.

The integration time was chosen to be between 30% and 40% of the vibration period. As reported in [10], this percentage is the optimal for avoiding a signal that is not sufficient due to a low integration time and for preventing excessive data averaging over a load cycle, which for a sinusoidal load means an amplitude of zero. All the acquisition parameters, including the alias value, are reported in Table 1.

Experimental Data Processing

The thermal sequences have been processed for evaluating the thermoelastic data in terms of amplitude and phase of the first harmonic component. Amplitude maps provide information on the distribution of elastic stresses and therefore elastic deformations induced by the mode of vibration. In effect, under adiabatic conditions the thermoelastic signal is proportional to the first stress invariant and can therefore be used to describe the vibration mode of the component [3].

In this regard, the IRTA2® software has been adopted to perform the signal reconstruction analysis by means of the following temperature model [7]:

$$T(t) = T_0 + bt + \Delta T_1 \sin(\omega t + \varphi_{1\omega}) + \Delta T_2 \sin(2\omega t + \varphi_{2\omega})$$

where T_0 is the mean temperature, b is the mean slope temperature increasing, ΔT_1 is the amplitude of the thermoelastic signal, ΔT_2 is the amplitude of the second harmonic temperature component, and $\varphi_{1\omega}$ and $\varphi_{2\omega}$ are the phase signals referred to the first and second harmonic components, respectively.

The output of the analysis is 2D maps where each pixel represents the specific feature. In particular, the present research is focused only on the analysis of ΔT_1 and $\varphi_{1\omega}$, corresponding to harmonic term of the excitation. The phase at each pixel was employed to determine the sign of the amplitude for illustration purposes. Assuming adiabatic conditions, this harmonic term is directly related to the variation in the sum of the principal stresses, $\Delta\sigma_{kk}$:

$$\Delta T = -T_0 K \Delta\sigma_{kk}$$

where T_0 is the temperature at the unloaded state and K is the thermoelastic constant that depends on material properties like the density, the heat capacity at constant pressure, and the thermal expansion coefficient.

FEM Model

An FEM model on Abaqus was performed for each specimen. Four-noded shell elements, 5 mm thick, were employed with an approximate size of 3 mm, but with a refined mesh around the discontinuity to accurately reproduce the stress gradient. An aluminum with 2700 kg m^{-3} density, Young's modulus of 69 GPa, and Poisson's ratio of 0.3 was employed as a material. To simulate the experimental conditions, the lower edge of the model was constrained.

A frequency analysis was performed to determine the eigenvalues and eigenvectors, i.e., the natural frequencies and mode shapes of the model. The Lanczos eigensolver was employed for a maximum frequency of interest of 2000 Hz, in agreement with the capabilities of IR thermography for high frequencies shown in the literature [13].

According to the experimental conditions and the symmetry of the load during the tests, the bending modes were analyzed. As the natural frequency identification during the experiments was performed using single-output measurement, the FEM model was employed as a reference to identify the bending modes based on the frequency.

Results

Mode shapes are orthogonal and linearly independents and, hence, develop different stress distribution along the specimen. According to this, each mode is differently affected by a geometric discontinuity. Taking a variety of modes, the aim is to identify the distortion in the field due to the geometric discontinuity.

Prior to show the results, it is worth noting that due to the nature of the modal eigenproblem, mode shapes, which are the eigenvectors, are undetermined. Only a ratio between amplitude coordinates can be obtained, so mode shapes are usually

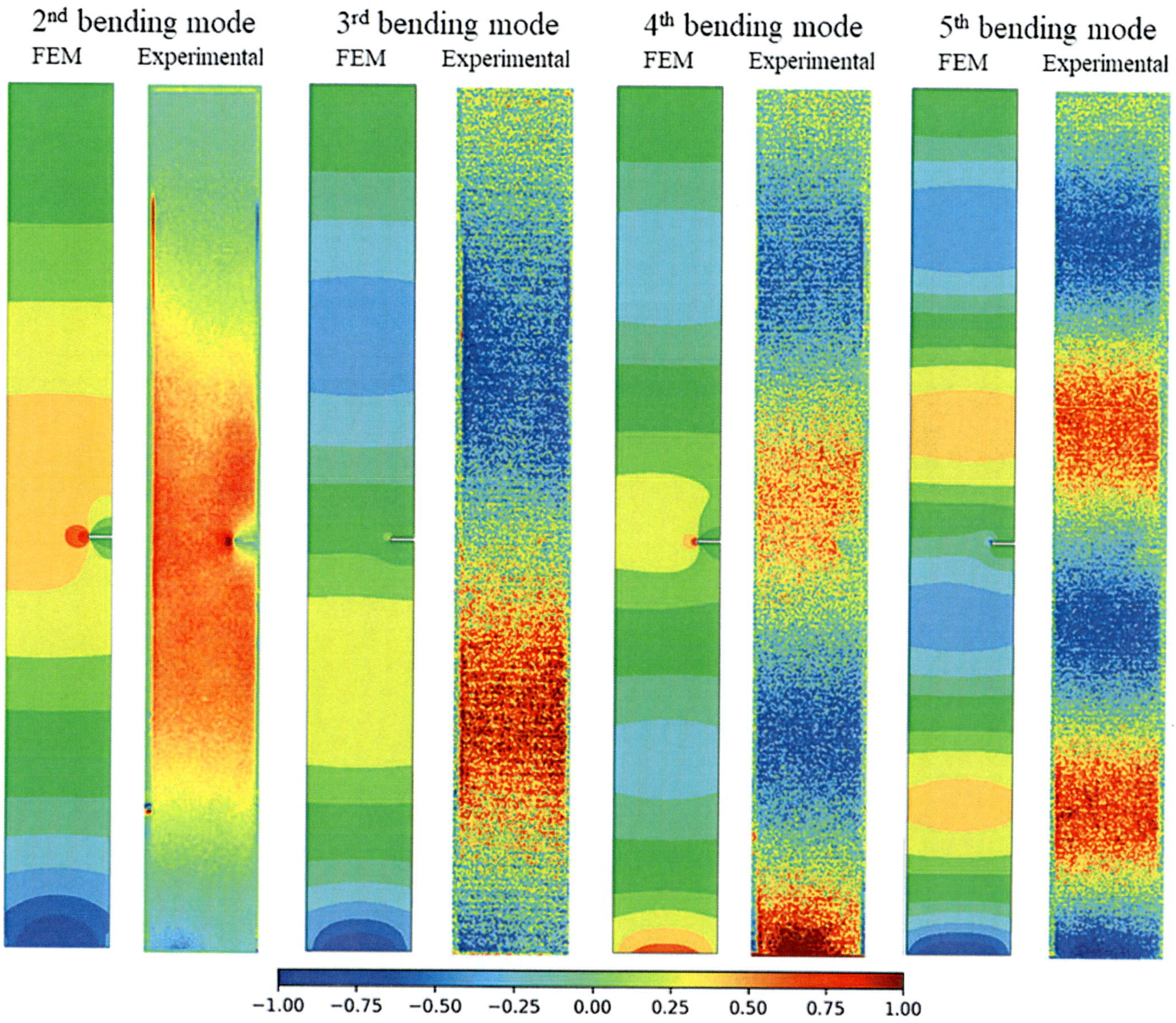

Fig. 4 Normalized first stress invariant field of the bending modes of the beam with a notch obtained by FEM and TSA

normalized dividing by the maximum value. Therefore, mode shapes do not represent real amplitude but shape. Since stresses are linearly related to displacements, here it is also convenient to use normalization.

Moreover, considering that the thermoelastic signal is proportionally related to the first invariant of stresses, the shape of the field is the same for temperature and stresses. Therefore, normalized temperature maps can be directly employed as stress maps without calibration.

In Fig. 4, the normalized maps of the sum of principal stresses are shown. Four bending modes were taken, comparing the FEM and experimental results. In FEM results, it is possible to see how the discontinuity can affect the field. Namely, those modes whose first stress invariant presents a high level around the discontinuity are more affected, as seen for the second and fourth bending modes. As can be observed, the experimental shapes show a good agreement with the FEM results. As demonstrated in previous studies, the amplitude of stresses diminishes with the mode order since the displacement also diminishes as the frequency of excitation rises. According to this, it is possible to observe that the second bending mode shows more clearly the distortion of the field due to the discontinuity. For the fourth mode, the noise level is higher and hides the stress gradient around the discontinuity when the size of the discontinuity in the image is small. The third and fifth bending modes have a node of stresses in that region, and the effect is subtle, but still visible. Observing the FEM results, the field is disturbed, and a small stress concentration in the upper corner of the notch can be identified. This is also observed experimentally in the fifth mode. For the third mode, it might be intuited but previous information would be necessary.

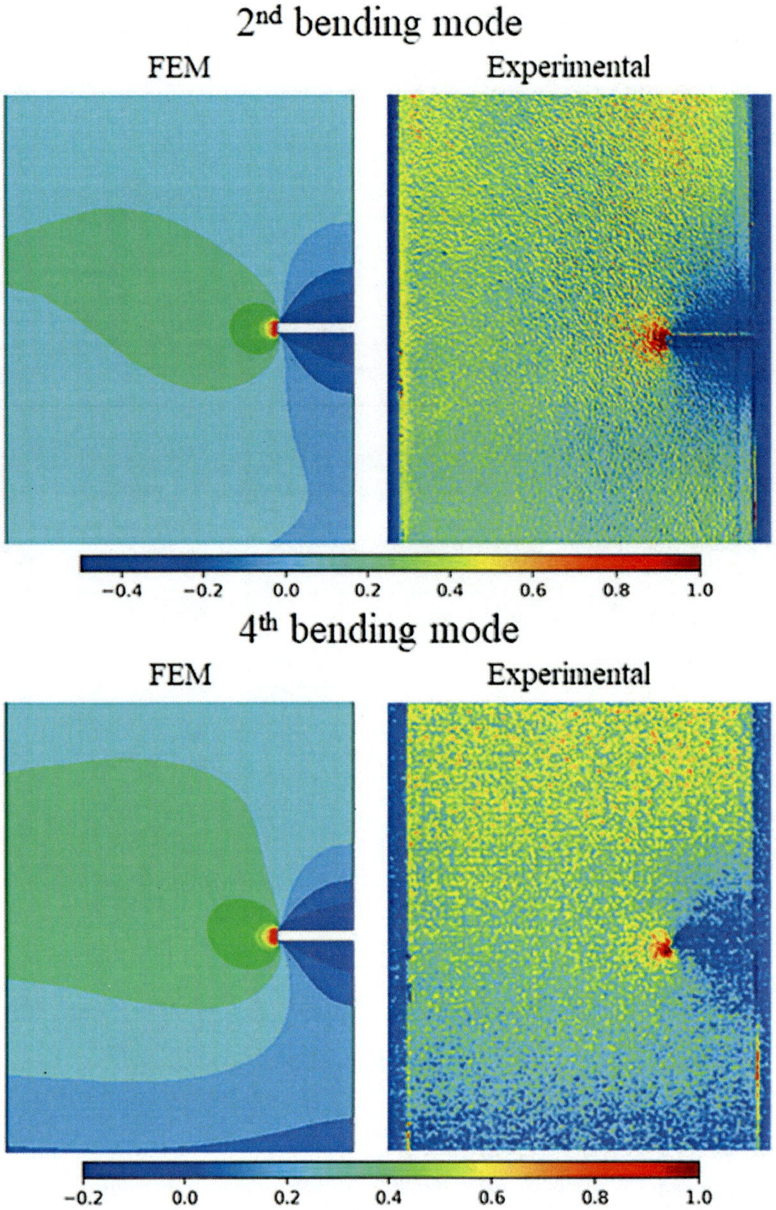

Fig. 5 Normalized first stress invariant field of the second and fourth bending modes of the beam with a notch obtained by FEM and TSA, employing a higher image resolution

In order to improve the resolution of the stress field, the camera was moved closer to the specimen to achieve a higher px-mm ratio in the discontinuity area. For this task, the second and fourth modes were analyzed for being the most sensitive to the discontinuity.

The results are shown in Fig. 5. As can be observed, this setup allows a better characterization of the gradients around the discontinuity. However, the noise is still relevant for the fourth mode. Despite increasing the px-mm ratio, which just improves the spatial resolution, the signal-to-noise ratio is expected not to vary as the level of the thermoelastic signal is the same. As mentioned before, the level of the signal decreases with frequency, so a Gaussian filter (size of 3 pixel and standard deviation of 2) was applied to the fourth mode maps to reduce the high-frequency noise.

Conclusion

This paper advances on the measurement of stress maps with infrared cameras due to thermoelastic effect under vibration, namely, on the characterization of sum of principal stresses occurring during resonance. Here, this approach is proposed to evaluate its capability to identify the distortion in the field by geometric discontinuities due to the interest of the stress concentration for structural integrity. The shape of the field was the aim of this study; thus, normalized maps from the thermoelastic signals were analyzed. Nevertheless, if real values of stress are required under a specific load, the thermoelastic constant can be determined.

An aluminum beam containing a notch was inspected through four resonances with bending deformation. The experimental results agreed with the FEM prediction. The presence and influence of the discontinuities were clearly observed for those resonances that have high stress in the surroundings of the discontinuity. Despite the noise typically present in the thermoelastic signals, especially under this kind of high-frequency excitation, it has been shown how increasing the image resolution a better description of the field around the discontinuity can be obtained. In sum, the results show that under the diversity of shapes offered by mode shapes, it is possible to identify the presence of a discontinuity, which may offer new approaches for the structural integrity evaluation.

Acknowledgments This work was supported by the Spanish Government through the program "José Castillejo" [grant number CAS21/00049]. This work was also part of a research project "MAIA - Infrastructure Active Monitoring" [project identifier: ARS01_00353], within the PON "RICERCA E INNOVAZIONE" 2014–2020 call supported by the Italian Government. The authors would like to thank Prof. L. Soria and Dr. S. De Carolis for the support given to the authors during the experimental work.

References

1. Dulieu-Barton, J.M., Stanley, P.: Development and applications of thermoelastic stress analysis. J. Strain Anal. Eng. Des. **33**(2), 93–104 (1998). https://doi.org/10.1243/0309324981512841
2. Pitarresi, G., Patterson, E.A.: A review of the general theory of thermoelastic stress analysis. J. Strain Anal. Eng. Des. **38**(5), 405–417 (2003). https://doi.org/10.1243/03093240360713469
3. Harwood, N., Cummings, W.M. (eds.): Thermoelastic Stress Analysis. IOP Publishing (Adam Hilger), Bristol (England) (1991)
4. Tighe, R.C., Dulieu-Barton, J.M., Quinn, S.: Identification of kissing defects in adhesive bonds using infrared thermography. Int. J. Adhes. Adhes. **64**, 168–178 (2016). https://doi.org/10.1016/j.ijadhadh.2015.10.018
5. Ancona, F., Palumbo, D., De Finis, R., Demelio, G.P., Galietti, U.: Automatic procedure for evaluating the Paris Law of martensitic and austenitic stainless steels by means of thermal methods. Eng. Fract. Mech. **163**, 206–219 (2016). https://doi.org/10.1016/j.engfracmech.2016.06.016
6. Allevi, G., Cibeca, M., Fioretti, R., Marsili, R., Montanini, R., Rossi, G.: Qualification of additively manufactured aerospace brackets: a comparison between thermoelastic stress analysis and theoretical results. Measurement. **126**, 252–258 (2018). https://doi.org/10.1016/j.measurement.2018.05.068
7. De Finis, R., Palumbo, D., Galietti, U.: A multianalysis thermography-based approach for fatigue and damage investigations of ASTM A182 F6NM steel at two stress ratios. Fatigue Fract. Eng. Mater. Struct. **42**(1), 267–283 (2019). https://doi.org/10.1111/ffe.12903
8. Galietti, U., Modugno, D., Spagnolo, L.: A novel signal processing method for TSA applications. Meas. Sci. Technol. **16**(11), 2251–2260 (2005). https://doi.org/10.1088/0957-0233/16/11/017
9. Di Renzo, A., Marsili, R., Martarelli, M., Moretti, M., Rosati, G., Rossi, G.L.: Simultaneous application of scanning laser vibrometry and thermoelasticity for measurement of stress-strain fields on mechanical components. In: Seventh International Conference on Vibration Measurements by Laser Techniques: Advances and Applications, June 2006, vol. 6345, p. 63450H. https://doi.org/10.1117/12.693161
10. Backman, D., Greene, R.J.: Gas turbine blade stress analysis and mode shape determination using thermoelastic methods. Appl. Mech. Mater. **13–14**, 281–287 (2008). https://doi.org/10.4028/www.scientific.net/AMM.13-14.281
11. Fruehmann, R.K., Dulieu-Barton, J.M., Quinn, S., Peton-Walter, J., Mousty, P.A.N.: The application of thermoelastic stress analysis to full-scale aerospace structures. J. Phys. Conf. Ser. **382**(1), 012058 (2012). https://doi.org/10.1088/1742-6596/382/1/012058
12. Phan, T.S., Dulieu-Barton, J.M., Temarel, P.: Thermoelastic stress analysis of structures under natural vibrations. Exp. Mech. **46**(4), 463–472 (2006). https://doi.org/10.1007/s11340-006-8445-6
13. Molina-Viedma, Á.J., Felipe-Sesé, L., López-Alba, E., Díaz, F.A.: Thermoelastic effect in modal shapes at high frequencies using infrared thermography. Measurement. **176**, 109180 (2021). https://doi.org/10.1016/j.measurement.2021.109180
14. Molina-Viedma, A.J., Felipe-Sesé, L., López-Alba, E., Díaz, F.A.: Comparison of lock-in correlation and a novel periodogram method for experimental multi-harmonic thermoelastic analysis. Mech. Syst. Signal Process. **164**(February 2021), 108235 (Feb. 2022). https://doi.org/10.1016/j.ymssp.2021.108235

Induction Thermography: Influence of Testing Parameters for Different Crack Geometry

Davide De Vanna, Ester D'Accardi, Giuseppe Dell'Avvocato, Davide Palumbo, and Umberto Galietti

Abstract In recent years, induction thermography has been proposed to detect and characterize shallow and open cracks on steel components. The specimen is inspected by an infrared camera that records the heating and cooling behavior after a short induced electrical current pulse. The presence of a crack changes the distribution of the induced eddy currents with a consequent more significant local heating around the region of interest than the related sound one. This work investigates the influence of the testing parameters with an experimental approach, considering the analysis of a ferromagnetic master specimen with imposed simulated open cracks of different depths. The influence of the relative position between the coil and simulated crack has been evaluated, considering both numerical and experimental results. Two different coils have been adopted to optimize the experimental tests in terms of inspection time and improve the signal-to-noise ratio. It has been demonstrated that the post-processing of the raw thermal data is necessary to have useful information about defect detection for a pulse duration of 10 ms with only 20 A. A linear relationship exists between the phase contrast and the nominal defect depth.

Keywords Induction thermography · Non-destructive testing (NDT) · Simulated cracks · Testing parameters · Post-processing algorithms

Introduction

Nowadays, induction thermography is a well-established technique for non-destructive testing of surface cracks in steel specimens and real components. During a typical induction test, the specimen to be inspected is placed inside or close to an induction coil. During and after applying a short heating pulse of typically 10–1000 ms, both currents and heat will be induced inside the inspected material, and an infrared camera records the thermal behavior [1–5]. Thanks to temperature evaluation and post-processing analyses, it is possible to detect surface cracks and in some cases also to characterize them in terms of sizes and depths [1].

However, different parameters need to be considered before carrying out the experimental tests, depending on the available equipment, the material, and the expected defect type and geometry. Different research works, mainly based on analytical and numerical approaches, studied these aspects [1–7], providing essential results for the correct application of this technique. In particular, there is a complete study by Prof. Oswald-Tranta [1] in which finite element simulations were carried out to investigate how phase contrast depends on parameters such as excitation frequency, pulse duration, material, crack depth, and inclination angle of the crack. In this way, an overview of how measurement parameters should be optimized for a given material to detect and characterize cracks has been obtained.

Essentially, the current flow is normally influenced by the crack geometry [1–5]; for this reason, there is a German standard DIN 54193:2018-02 [8] that provides some useful indications about the construction of master specimens to evaluate the influence of different defect geometric features in relation to an induction test. Furthermore, in the literature, the research works that propose using low-cost heating source induction system for components inspection are very few [9, 10]. On the other hand, several works propose its use for inspecting of real components and industrial applications [6–9, 11, 12].

D. De Vanna · E. D'Accardi · G. Dell'Avvocato · D. Palumbo (✉) · U. Galietti
Department of Mechanics, Mathematics and Management, Polytechnic University of Bari, Bari, Italy
e-mail: davide.devanna@poliba.it; ester.daccardi@poliba.it; giuseppe.dellavvocato@poliba.it; davide.palumbo@poliba.it; umberto.galietti@poliba.it

C. Franck et al. (eds.), *Challenges in Mechanics of Biological Systems and Materials, Thermomechanics and Infrared Imaging, Time Dependent Materials and Residual Stress, Volume 2*, Conference Proceedings of the Society for Experimental Mechanics Series,
https://doi.org/10.1007/978-3-031-50470-9_11

In this work, starting from the analysis of a first ferromagnetic master specimen [13–15] with open cracks of different superficial depths, the influence of relative position between coil and crack was investigated, considering a typical solenoid circular coil and a very low-cost heating source, that is, a ZVS (zero voltage switching) induction heater of just 1000 W. Obviously, the influence of defect depth was also evaluated and discussed. Some preliminary numerical analyses were carried out on COMSOL Multiphysics® before starting the experimental campaign, to understand the thermal behavior along the entire crack length and at the crack tip after an induction test, with the same constraints and technical set-up specifications.

Moreover, to reduce the testing time and make the technique more suitable for an industrial application (inspecting larger areas at the same time), a different coil geometry has been adopted, with a rectangular shape and a main preferential direction for the current flow lines. The obtained results have demonstrated an influence of the relative position between the coil and imposed crack and that the phase contrast has a linear dependence on the defect superficial depth. Besides, it has also been demonstrated that a pulse duration of 10 ms is sufficient to detect the imposed defects, especially in the case of surface inspection with a rectangular coil, if a post-processing analysis is performed to evaluate the raw thermal data.

Theory

Induction thermography's main physical and theoretical background involves induced eddy current and thermal heating transfer. Considering a one-dimensional solution, after an induction test that starts at time $t = 0$ with a finite pulse duration indicated as t_{risc}, it is possible to obtain the related temperature increase due to the Joule effect of the induced currents in the sound area of the material from the equation of Carslaw and Jaeger [16] reported below:

$$\Delta T_{\text{sound}} = \frac{B_{\text{tot}}^2}{2\mu_0}\mu_0\mu_r 2\sqrt{\pi} f \delta \frac{\sqrt{t_{\text{risc}}}}{\sqrt{\lambda\rho c_p}}$$

where B_{tot} $[T]$ is the magnetic field within the considered point [17, 18], μ_0 [H/m] is the magnetic vacuum, μ_r [H/m] is the magnetic permeability related to the inspected component, f [Hz] is the excitation frequency, t_{risc} [ms] is the pulse duration during the heating of the component, λ [W/mk] is the thermal conductivity of the material, ρ [Kg/m] is the material density, and c_p [J/Kg*K] is the specific heat.

Starting from the evaluation of the temperature increase in the sound area, it is possible to estimate the thermal behavior of the crack as additional heating with respect to the sound. In particular, the crack is represented as a discontinuity of an infinite length a finite depth d, with a perpendicular direction to the inspected surface:

$$\Delta T_{\text{crack}} = \Delta T_{\text{sound}}\left(1 + \frac{1}{\sqrt{\pi}}\frac{d}{\mu_{th}}E_i\left(\frac{d^2}{\mu_{th}^2}\right) + \text{erf}\left(\frac{d}{\mu_{th}}\right)\right)$$

where E_i is the exponential integral function and the erf is the error function.

With the previous equations (one-dimensional models), it is possible to obtain the heating trends for a sound and a defective area. The equation related to the cooling phase is instead reported below, in the case of the analysis of a sound area, easily replaceable in the previous equation to obtain an estimation of the cooling trend in the case of a crack.

$$\Delta T_{\text{sound}} = \frac{B_{\text{tot}}^2}{2\mu_0}\mu_0\mu_r 2\sqrt{\pi} f \delta \frac{\sqrt{t} - \sqrt{t - t_{\text{risc}}}}{\sqrt{\lambda\rho c_p}}$$

Experimental Set-up

Following the German standard DIN 54193:2018-02, a ferromagnetic steel S/235 specimen with 12 defects at different depths from the surface has been considered. The details of the specimen and investigated defects geometry are reported in Table 1. The used set-up is shown in Fig. 1. A cooled thermal sensor FLIR A6751 acquired the thermal sequences, while the specimen has been stimulated with a portable solution ZVS induction heating card, controlled by Arduino. Technical set-up specifications are reported in Table 2. The pulse duration has been changed three times to study the influence of this parameter.

Table 1 Specimen and defect geometry details

Specimen geometry				
Length [mm]	Width [mm]	Thickness [mm]		
200	60	3		
Defect geometry				
Depth [mm]	Depth step [mm]	Length [mm]	Thickness [mm]	Distance (between two consecutive defects) [mm]
0.15 – defect 1	0.25	10	0.4	0.15

(a) (b)

Fig. 1 Experimental campaign – induction thermography, (**a**) specimen geometry and (**b**) set-up (ZVS induction heating card – solenoid circular coil 60 mm; maximum power available 1000 W, current up to 20 A–480 W, supply 12–48 V) – cooled thermal sensor FLIR A6751)

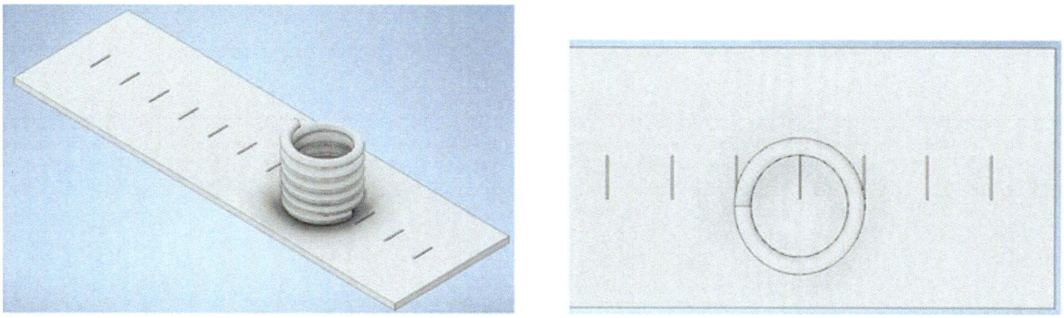

(a) (b)

Fig. 2 A scheme of the relative position between the coil and the inspected specimen surface, (**a**) lateral and (**b**) top view

The relative position between simulated cracks and the coil is reported as a scheme in Fig. 2. To inspect a larger surface area at the same time, some experimental tests have been carried out with a rectangular coil with a smaller width equal to the radius of the adopted circular coil, a length of about 90 mm, and the same turn number. Three replications of the same test have been carried out (Table 2).

Analysis

As the first and preliminary approach to understanding the influence of the relative position of a circular coil and the crack and the influence of the geometric characteristics of the investigated specimen, a FEM model was developed adopting COMSOL

Table 2 Testing parameters for the excitation source (portable system ZVS inductor) and the thermal cooled camera

Inductor ZVS parameters					
Current [A]	Coil diameter [mm]	Power (W)	Distance coil – specimen [mm]	Excitation frequency [kHz]	Pulse duration [ms]
20	40	480	5	110–120	10–100–500
Thermal camera parameters FLIR A6751					
Integration time [ms]	Calibration range [°C]	Frame rate [Hz]	Time of acquisition [s]	Window [pixels]	Spatial resolution [mm/pixel]
0.9741	10.0–90.0	900	8	160 × 120	0.12

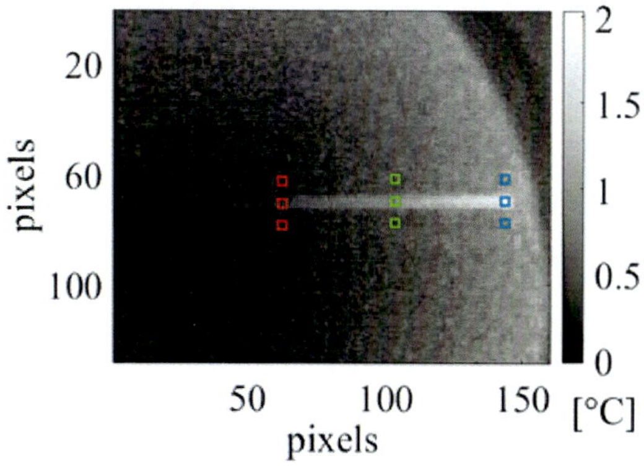

Fig. 3 Scheme of the positions of sound and defective areas along the entire length of the simulated crack. Thermal map after a pulse excitation of 500 ms (end of heating), simulated crack with a length of 10 mm, and a depth of 1.15 mm

Multiphysics®. The FEM model considers the Electromagnetic Heating Multiphysics module, which allows coupling the physics related to electromagnetic induction, with the one related to heat transfer. Within this module the physical parameters of the inductor have been defined such as the excitation frequency, the current value, and the current flow direction. Finally, a mesh was generated and characterized by more elements in correspondence with the simulated crack. A free tetrahedral mesh was adopted with a maximum element size in the proximity of the crack of 200 μm and a minimum element size of 50 μm. Other useful details for the mesh are specified below, such as a maximum element growth rate of 1.5, a curvature factor of 0.2, and a number of elements for the entire geometry of about 65817.

For the analyses of the acquired thermal sequences, it was necessary to consider the raw thermal data and extract suitable thermal features for defect characterization. First, the sound and defective area were defined, considering the total length of the simulated crack and the relative position between the coil and crack. A scheme of the adopted approach is depicted in Fig. 3, considering three different defective areas and the related sound (regions of interest – RoIs).

The apparent maximum temperature and the contrast were used as the main thermal features to discuss the obtained results; furthermore, simple post-processing has been carried out to extract some thermal features, considering, as main algorithms, the pulsed phase thermography (PPT) and the principal component thermography (PCT) [1–3, 11, 19–22]. The same post-processing analyses have been repeated several times for all the replications considering the heating phase and part of the cooling phase as suggested in the literature and schematically shown in Fig. 5.

Before applying these algorithms, the delta temperature was obtained by subtracting the first cold frame from the entire sequence of raw thermal data.

Numerical Results

Figures 4 and 5 show some numerical simulation results after a pulse heating of 500 ms for a given relative position between the coil and the simulated crack. In particular, in Fig. 4 the superficial current density [A/m^2] is reported, considering the behavior along the entire crack length in two different views. These results demonstrated greater concentration of the induced current at the crack tip. With the aim to highlight the different behavior along the length of the simulated crack in Fig. 5 is reported the thermal behavior, considering the delta temperature with two different scales.

A simulation was also carried out to evaluate the effect of the edges of the specimen. To do this, the closest crack to the edge of the specimen has been considered (about 15 mm). The obtained results demonstrate that the currents tend to thicken around the edge specimen's edge, as shown in Fig. 6. As expected, a significant temperature increase can be observed at the edges after a pulse duration of 500 ms. It is worth noticing that this effect could hide the actual one due to the presence of the crack.

Experimental Results

In Fig. 7 some results are reported, as an example, considering the test with a pulse duration of 500 ms. Three different frames and the map of the maximum delta temperature calculated pixel by pixel considering the entire sequence after the cold frame subtraction are reported. From these thermal maps, due to the small amount of available current and relative position between the coil and the simulated defect, it is possible to note a different thermal behavior along the entire length

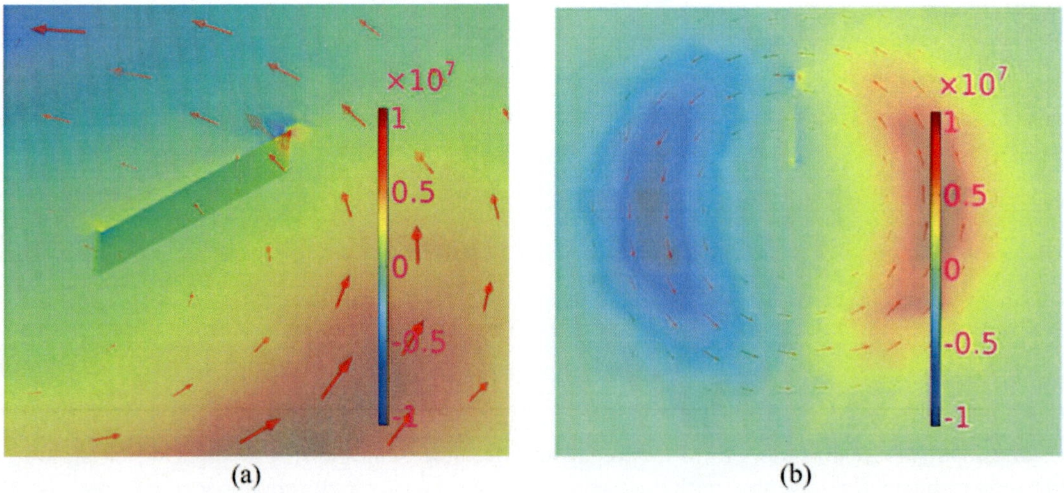

Fig. 4 Eddy current density – y component along the entire simulated crack length, after an induction heating of 500 ms – numerical simulation; (**a**) side view and (**b**) top view after the heating phase

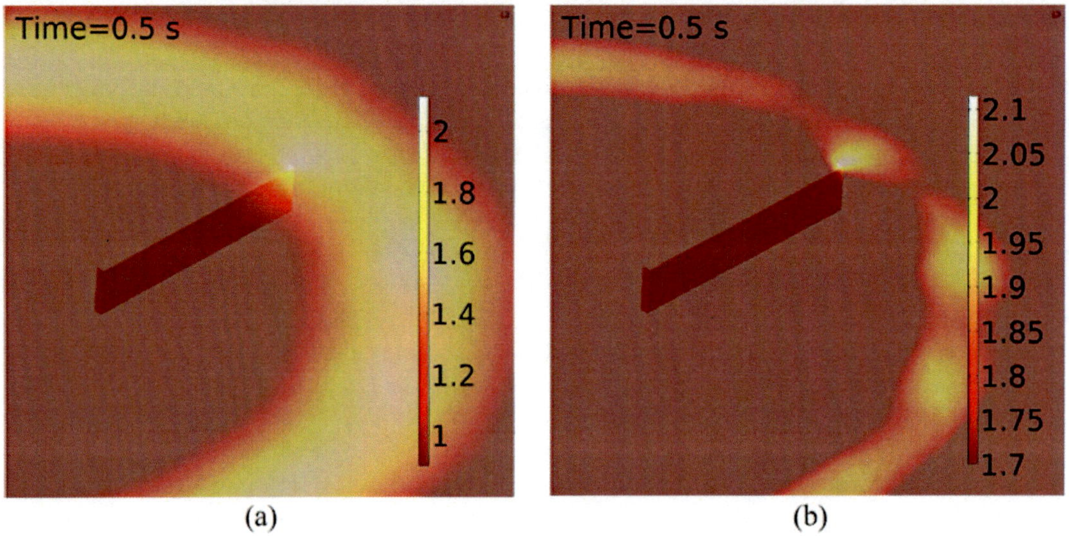

Fig. 5 Surface temperature increase after a numerical simulation of an induction heating of 500 ms; (**a**) delta temperature at the end of the heating along the entire crack length and (**b**) a different scale to highlight the thermal behavior near the crack tip

Fig. 6 Edge effect that influences the current flow with a consequent decrease of the temperature increase at the crack tip – delta temperature after a pulse duration of 500 ms

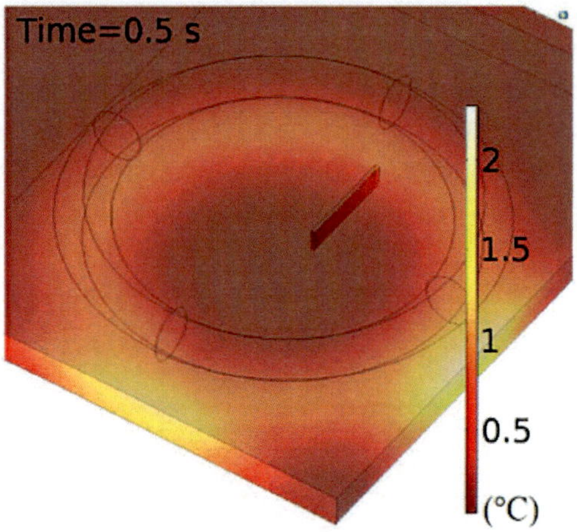

of the crack: the different points reach the maximum temperature at different times. Due to the induced currents, the Joule effect interests only the closest part of the simulated crack near the coil. Then, heat spreads along the crack during the cooling

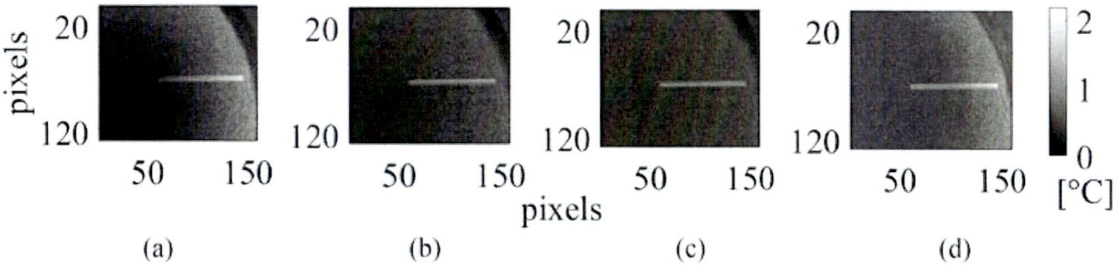

Fig. 7 Thermal maps related to a sequence of raw thermal data for a pulse excitation of 500 ms: (**a**) map of the maximum delta temperature (500 ms), (**b**) cooling phase after 1 s of analysis, (**c**) cooling phase after 2 s of analysis, and (**d**) map of the maximum delta temperature along the entire sequence pixel by pixel – simulated crack with a length of 10 mm and a depth of 1.15 mm

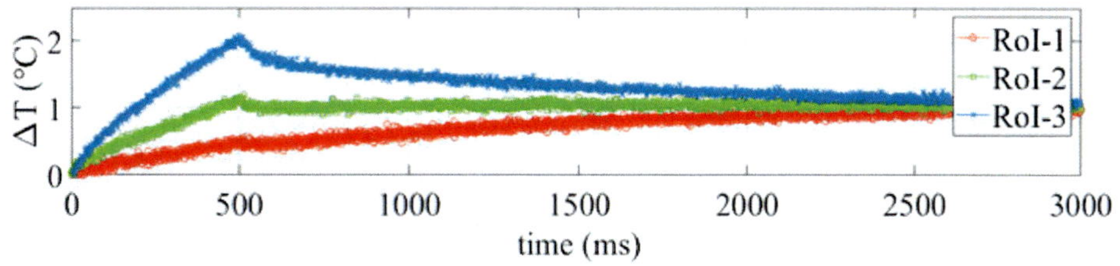

Fig. 8 Temperature increases after a pulse duration of 500 ms in three different regions of interest along the entire length of the simulated open crack – simulated crack with a length of 10 mm and a depth of 1.15 mm

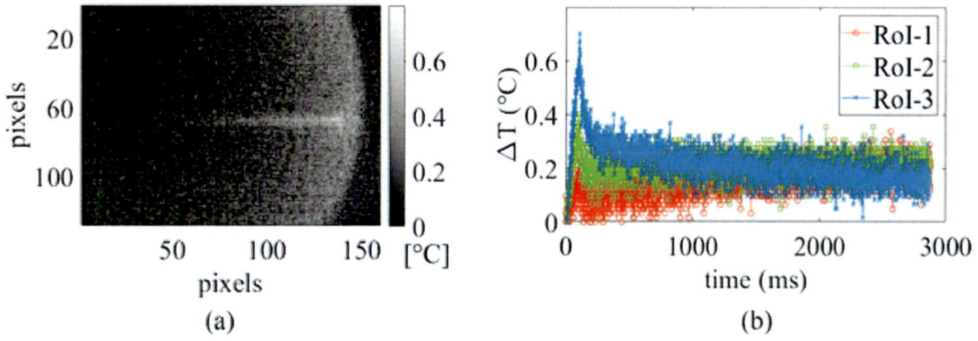

Fig. 9 Results of the induction test with a pulse duration of 100 ms: (**a**) raw thermal map taken at the end of the heating phase and (**b**) trends related to the regions of interest chosen to observe the thermal phenomena – simulated crack with a length of 10 mm and a depth of 1.15 mm

due to the thermal diffusion effect. As confirmation of these results, the trends related to the three different regions of interest highlighted before are reported in Fig. 8. This behavior is typical of an open discontinuity and very different from a closed crack.

Figures 9 and 10 show the obtained results considering two different pulse durations, respectively, of 100 and 10 ms. In particular, Figs. 9a and 10a show raw thermal maps corresponding to the end of the heating period; instead Figs. 9b and 10b show the trends of the investigated regions of interest along the crack length. Especially in the case of a pulse duration of only 10 ms, it is hard to distinguish the presence of a defect, because the signal is very noisy.

Among the main post-processing algorithms used in literature to analyze raw thermal data, pulsed phase thermography is the oldest and most common. In the literature, there is no precise indication about frame rate and number of frames to adopt for induction tests. Usually, the analysis is performed considering both the heating and the cooling phase. As a single map used to represent the obtained results, the second frequency of the entire spectrum has been selected, corresponding to the maximum phase contrast localized around the crack tip.

Another widespread post-processing algorithm is principal component thermography (PCT). As known, the main result of this type of post-processing is related to the extraction of the empirical orthogonal functions, principally the first five or six components.

As an example of the obtained results, Figs. 11 and 12 show the results for both algorithms, considering a test of 500 ms with a number of data equal to the entire heating phase and a part of the cooling with the exact duration of the heating pulse (Fig. 5).

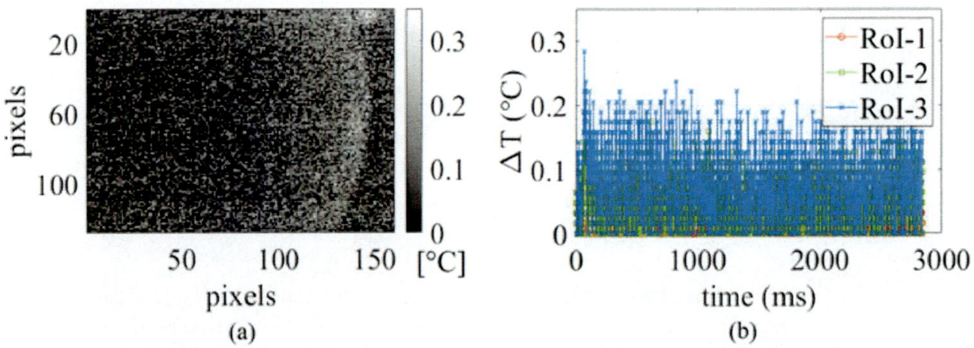

Fig. 10 Results of the induction test with a pulse duration of 10 ms: (**a**) raw thermal map taken at the end of the heating phase and (**b**) trends related to the regions of interest chosen to observe the thermal phenomena – simulated crack with a length of 10 mm and a depth of 1.15 mm

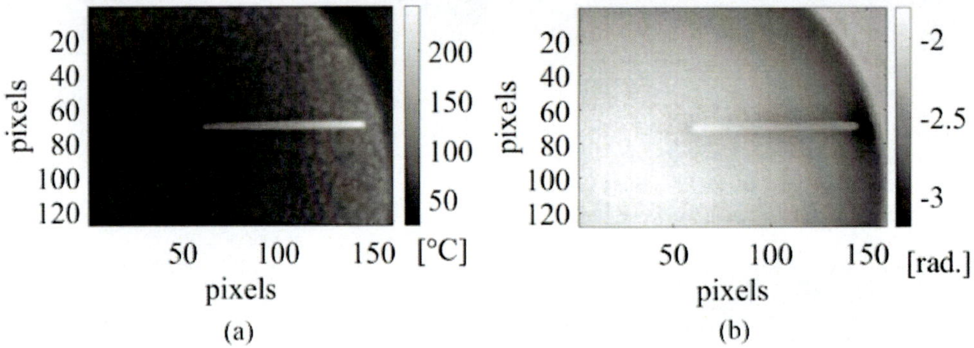

Fig. 11 (**a**) Amplitude and (**b**) phase results after the application of PPT algorithm considering a number of frames equal to twice the heating phase – frequency of 1 Hz

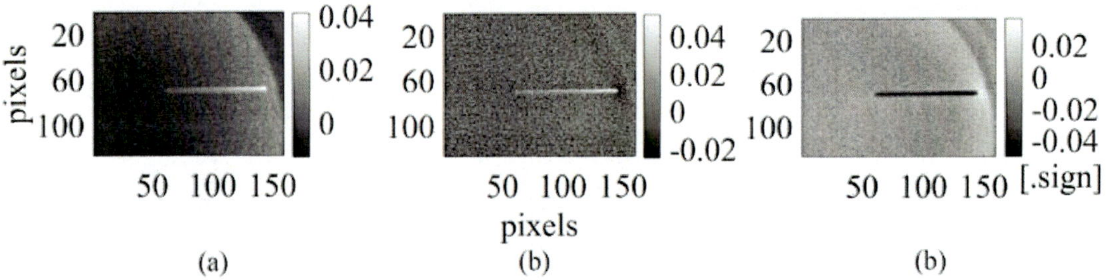

Fig. 12 Results after the application of the principal component analysis: (**a**) EOF 2, (**b**) EOF 3, (**c**) EOF 4 – pulse duration 500 ms, simulated crack with a length of 10 mm, and a depth of 1.15 mm

A similar analysis was carried out considering a pulse duration of 10 ms. The results obtained after applying the principal component thermography are reported in Fig. 13 (raw thermal data in Fig. 10).

Discussion

The post-processing analyses have been repeated at different times for all the replications and imposed cracks. Interesting results emerge from the analysis because there is a linear correlation between the phase contrast and the defect depth (Fig. 14a). Defects near the edge of the specimen present a phase contrast that slightly deviates from the linear trend. In fact, as demonstrated before, by the numerical simulation results with the same technical constraints (Fig. 6), the currents tend to thicken around the edge of the specimen. For this reason, these points were removed from Fig. 14a reporting the same results in Fig. 14b, with a more linear trend.

For the test with a pulse duration of 10 ms, after different attempts, it is impossible carry out a quantitative analysis just from the analysis of raw thermal data due to the very noisy signal, due to the current value of just 20 A. For this reason,

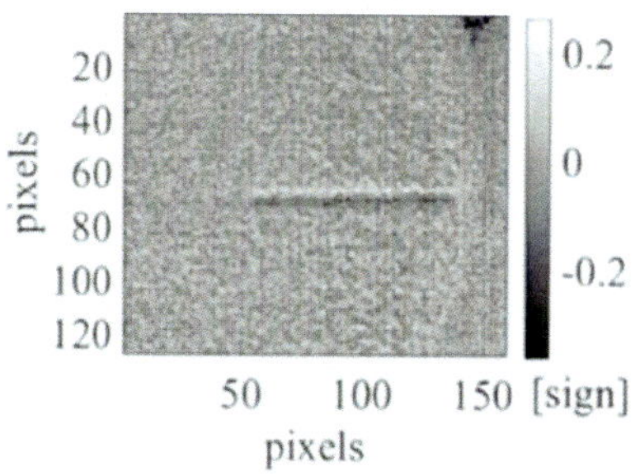

Fig. 13 Result related to the principal component analysis and to the EOF 4 after an induction test of 10 ms for the simulated crack of a length equal to 10 mm and with a depth of 1.15 mm

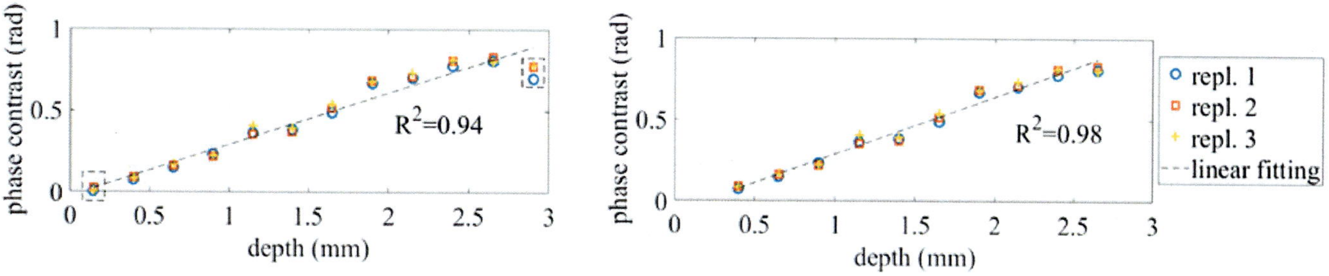

Fig. 14 Phase contrast results for all the imposed cracks after an induction test of 500 ms, analyzing both heating and cooling phase together (RoI in proximity of the simulated crack tip) – (**a**) all the imposed defects are considered; (**b**) the edge effects are included, removing the influenced points (first and last defect)

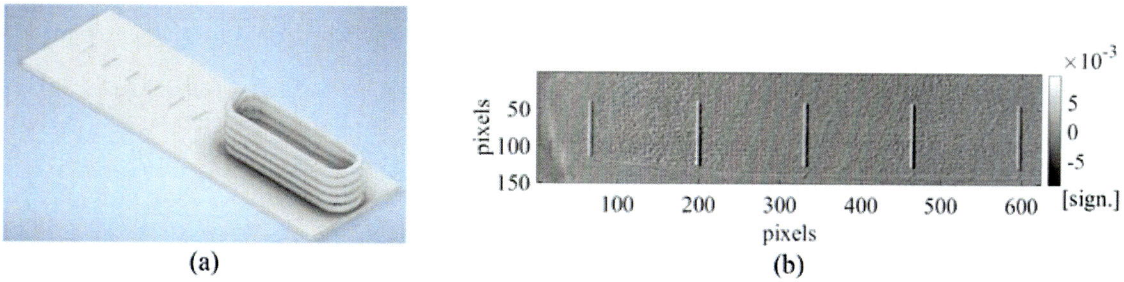

Fig. 15 An example of the obtained results considering a coil with a rectangular shape and a pulse duration of 10 ms: (**a**) exemplificative scheme of the adopted set-up and (**b**) preliminary result after a post-processing analysis – principal component analysis EOF4

in order to improve the obtained results, some preliminary tests have been carried out on the same specimen, with a coil of rectangular shape. The adopted coil geometry allows the simultaneous excitation of five different simulated cracks, offering a first advantage in technical times for component inspection. A prominent and preferential geometric dimension of the coil influences the induced current flow. In this case, the induced current flow will show a rectilinear trend, parallel to the main direction of the coil. In this way, it is possible for all five inspected cracks, to find the optimal condition in which the induced currents are perpendicular to the crack. Finally, compared to the circular one, the coil width guarantees a higher current density in the area enclosed by the coil profile, considering the same test parameters (scheme in Fig. 15a). Figure 15b shows an example of the obtained result after an induction test of 10 ms, considering as a post-processing algorithm the principal component analysis with the same approach explained before. After a qualitative comparison with the previous results (Fig. 13), this map shows a higher signal-to-noise ratio, improving defect detection even at very low pulse durations.

Conclusions

This work shows some preliminary results of an extensive numerical and experimental campaign on master specimens with imposed cracks, starting from the technical specifications suggested inside the reference DIN 54193:2018-02. In this first study, considering different imposed open cracks with a finite length and different depths, the influence of the coil position was evaluated. Considering a typical solenoid circular coil and positioning the turns near the crack tip, the obtained results were discussed with both a numerical and an experimental approach. The presence of an edge near the crack changes the trend of the current lines with a consequent decrease in the thermal contrast at the crack tip. Three different pulse durations were evaluated, up to a minimum of 10 ms. After a post-processing analysis, and considering a coil shape optimization, and a very low-cost set-up with a current value of only 20 A and 10 ms as pulse duration allows us the defect detection, with an inspection area of approximately 90 mm. This set-up and used approach could be helpful for industrial applications and analyses of actual components, considering all on a larger scale, starting from a higher current value and increasing the main dimension of the rectangular coil (based on the dimensions of the inspected components). As experimental phenomena to be further investigated, the defect depth shows a linear influence on the achieved phase contrast at the defect crack tip, with a consequent potential in the quantitative characterization of surface defects.

Acknowledgments The authors want to thank the Tesmec Rail company and in particular Ing. Gianluca Fumarola and Ing. Gaetano Marinelli for providing the master specimen investigated in this work useful for the joint research agreement commissioned at the Polytechnic University of Bari for the "Diagnostics of the rail integrity with innovative non-destructive test methods."

References

1. Oswald-Tranta, B.: Induction thermography for surface crack detection and depth determination. Appl. Sci. **8**(2), 257 (2018)
2. Mendioroz, A., Fuggiano, L., Venegas, P., de Ocáriz, I.S., Galietti, U., Salazar, A.: Characterizing subsurface rectangular tilted heat sources using inductive thermography. Appl. Sci. **10**(16), 5444 (2020)
3. Oswald-Tranta, B.: Time-resolved evaluation of inductive pulse heating measurements. Quant. InfraRed Thermogr. J. **6**(1), 3–19 (2009)
4. Netzelmann, U., Walle, G., Lugin, S., Ehlen, A., Bessert, S., Valeske, B.: Induction thermography: principle, applications and first steps towards standardisation. Quant. InfraRed Thermogr. J. **13**(2), 170–181 (2016)
5. Netzelmann, U.: Induction thermography of surface defects. In: Handbook of Advanced Nondestructive Evaluation, pp. 1497–1522. Springer, Cham (2019)
6. Yang, R., He, Y., Gao, B., Tian, G.Y., Peng, J.: Lateral heat conduction based eddy current thermography for detection of parallel cracks and rail tread oblique cracks. Measurement. **66**, 54–61 (2015)
7. Vrana, J., Goldammer, M.: Induction and conduction thermography: from the basics to application. Thermografie-Anwenderkonferenz, Munich (2017)
8. DIN 54193:2018-02
9. Toasa Caiza, P.D., Schwendemann, R., Calero, P., Ummenhofer, T.: Portable generator to detect cracks on large steel structures: an application of inductive thermography. J. Nondestruct. Eval. **40**(3), 63 (2021)
10. WO2018189209A1 – Method and devices for contactlessly and directly heating liquids and solids
11. D'Accardi, E., Palumbo, D., De Vanna, D., de Panizza, G., Galietti, U.: Evaluation of superficial cracks as real defects in railway applications by means of active thermographic techniques. In: 16th Quantitative InfraRed Thermography Conference, Paris, France, July 2022
12. Tuschl, C., Oswald-Tranta, B., Eck, S.: Scanning inductive thermographic surface defect inspection of long flat or curved work-pieces using rectification targets. Appl. Sci. **12**(12), 5851 (2022)
13. Walle, G., Netzelmann, U.: Thermographic crack detection in ferritic steel components using inductive heating. Proc. 9th ECNDT Berlin, 25(29.9) (2006)
14. Jäckel, P., Netzelmann, U.: The influence of external magnetic fields on crack contrast in magnetic steel detected by induction thermography. Quant. InfraRed Thermogr. J. **10**(2), 237–247 (2013)
15. Oswald-Tranta, B.: Thermoinductive investigations of magnetic materials for surface cracks. Quant. InfraRed Thermogr. J. ***1***(1), 33–46 (2004)
16. Carslaw, H.S., Jaeger, J.C.: Conduction of Heat in Solids. Clarendon (1959)
17. Callaghan, E.E.: The Magnetic Field of a Finite Solenoid, vol. 465. National Aeronautics and Space Administration (1960)
18. Derby, N., Olbert, S.: Cylindrical magnets and ideal solenoids. Am. J. Phys. **78**(3), 229–235 (2010)
19. Maldague, X.: Infrared and Thermal Testing, Nondestructive Testing Handbook, Vol. 3. ASNT, Columbus (2001). Available online: http://www.asnt.org. Accessed on 1 Jan 2018
20. Ibarra-Castanedo, C., Maldague, X.: Pulsed phase thermography reviewed. Quant. Infrared Thermogr. J. **1**(1), 47–70 (2004)
21. Rajic, N.: Principal Component Thermography. Defence Science and Technology Organisation Victoria (Australia) Aeronautical and Maritime Research Lab (2002)
22. D'Accardi, E., Palumbo, D., Galietti, U.: Experimental procedure to assess depth and size of defects with pulsed thermography. J. Nondestruct. Eval. **41**(2), 41 (2022)

Numerical Simulation of the Heat Dissipation During the Fatigue Test

Mohammad Zaeimi, Rosa De Finis, Davide Palumbo, and Umberto Galietti

Abstract Fatigue is an irreversible process accompanied by the heat dissipation which is significant when the transition from anelastic to inelastic strains happens. In view of this, in the last years, the heat dissipation has been accepted as an appropriate damage indicator of the material.

The estimation of the heat dissipation can be obtained by detecting the surface thermal footprint of the specimen by using thermography-based techniques. However, the energy dissipation as heat is highly sensitive to the environmental and test conditions and the microstructure status. Therefore, the experimental measurement is always associated with some inaccuracies and only provides an estimation of the heat dissipated during fatigue.

This paper is mainly focused on the numerical modeling of the heat dissipation performed by COMSOL Multiphysics software in order to investigate the factors that can affect the estimation of the heat source by means of thermography. The obtained results have been compared with an analytical solution derived from the one-dimensional heat equation. This study can provide valuable insights about the shape of the heat sources produced during the cyclic loading and differences associated with thermographic measurements and actual values, which are the main goals of this work.

Keywords Heat dissipation · Fatigue test · Thermography technique · Finite element simulation

Introduction

The study of fatigue phenomena remains an open issue because of the difficulties in modeling the material behavior and represents a key feature in choosing the most appropriate material since most of the mechanical components are subjected to dynamic loadings during their life [1].

Fatigue damage is an ambiguous phenomenon that forms and propagates mostly on a microscopic scale. Only when the damage has reached a critical level, it becomes detectable. In effect, by imposing a cyclic loading (depending on the loading level), the formation of dislocations is prompted. The accumulation of dislocations leads to plastic deformation. In turn, the occurrence and accumulation of microcracks are involved in macrocracks [2, 3]. Finally, the propagation of macrocracks induces material failure [4].

More insights on the phenomenon have been provided by Mareau et al. [5]. In particular, Mareau discussed two mechanisms of heat dissipations that occur during a fatigue test: (i) for low-stress amplitudes, an anelastic mechanism where the strains are irreversible and recoverable (e.g., the motion of dislocations, the migration of atoms, dislocation relaxation phenomena, and so on) and (ii) for high-stress amplitudes, an inelastic mechanism where the strains are irreversible but unrecoverable (e.g., the viscoplastic slip of dislocations). The transition from these two is the main cause of the strong dissipation and consequently the failure.

M. Zaeimi (✉) · D. Palumbo · U. Galietti
Department of Mechanics, Mathematics and Management, Polytechnic University of Bari, Bari, Italy
e-mail: mohammad.zaeimi@poliba.it; davide.palumbo@poliba.it; umberto.galietti@poliba.it

R. De Finis
Department of Engineering for Innovation, University of Salento, Lecce, Italy
e-mail: rosa.definis@unisalento.it

C. Franck et al. (eds.), *Challenges in Mechanics of Biological Systems and Materials, Thermomechanics and Infrared Imaging, Time Dependent Materials and Residual Stress, Volume 2*, Conference Proceedings of the Society for Experimental Mechanics Series,
https://doi.org/10.1007/978-3-031-50470-9_12

Of course, what happens to microscopic level determines an energy dissipation and a sentinel of this energy can be the surface temperature detected during a fatigue test. In this way, the aim of the present work is to investigate the capability of detected surface temperature to represent the temperature of inner heat source related to the fatigue damage.

One of the ways to take advantage of experimental techniques, particularly thermography, is to use them to detect the signature of materials undergoing to fatigue processes. In literature, different methods relying on the assessment of thermal signature have been developed to study the various aspects of fatigue damage in a rapid and reliable way [2, 6–8].

The first thermography-based method was the one proposed by Risitano [8] where a rapid method to study fatigue behavior of materials by assessing the temperature at stabilization of the material [8, 9] was investigated. In recent years, other different thermal features were investigated [10–23].

Enke and Sandor [10] for the first time showed the possibility of using the heat dissipation in the assessment of the damage of material. However, the quantification of the intrinsic dissipation, as an appropriate index for fatigue characterization, is a big challenge due to the fact that it is significantly dependent on the accuracy of temperature acquisition process and simplifications applied on heat equation for deriving it; therefore, a certain deviation always exists for these estimations [24–30]. Besides the estimations based on the mean temperature [31, 32], two other effective ways to estimate the heat dissipation are both studying the second amplitude harmonic (SAH) of temperature [10, 21, 33, 34] and the plastic work (hysteresis loop) [22], as they are completely related to the plastic deformation which is the main cause of the dissipation. Enke and Sandor [9] found the SAH effect by imposing the Fourier sine series on the thermal signal and proposed that the amplitude at the frequency twice the frequency of the mechanical loading is proportional to the plastic strains or intrinsic dissipation [34, 35].

In this study, a numerical model was defined to compare the heat dissipation experimentally estimated by measuring surface temperature and the numerical one using a FEM model that simulates the heat source. The effect of the amplitude and size of the heat source on the simulated temperature is also studied. The difference between measured and simulated temperature and energy values are also showed for different case studies. The present study could represent a step forward in understanding the difference between energy dissipations from the real heat source and energy dissipations assessed via thermography. It should be noted that the experimental values are taken from work by De Finis et al. [21].

The paper is organized as follows. First, the numerical model is proposed, and its correctness is shown by comparing the results with an analytical solution. Next, it is proposed how a heat source can be estimated in terms of magnitude and shape; and finally, the conclusion and future works are presented.

Numerical Simulation of the Heat Source

One way to ensure about the accuracy of a numerical model, especially in the case of cyclic heat source, is searching for a model that can provide an analytical solution. Generally, for an isotropic and homogeneous material, the first law of thermodynamics applied to a control volume surrounding the gage section of a sample in terms of energy per second (power) gives:

$$\underbrace{\int\left(\rho C\frac{\partial T}{\partial t}\right)dV}_{\text{Internal energy}} = \left[\underbrace{\int\left(k\nabla^2 T\right)dV}_{\text{Conduction}} - \underbrace{\int h\left(T - T_\infty\right)dA}_{\text{Convection}} - \underbrace{\int \epsilon\beta\left(T^4 - T_\infty^4\right)dA}_{\text{Radiation}}\right] + \underbrace{\int(\sigma : d\varepsilon)\, f\, dV}_{\text{Mechanical energy}} \quad (1)$$

where ρ, C, T, and T_∞ are correspondingly the density, specific heat at constant pressure, and absolute and ambient temperatures; h, k, ϵ, and β are, respectively, the convection coefficient, thermal conductivity, emissivity, and Stephan-Boltzmann constant; A, V, σ, ε, and f are, respectively, the area, volume of the gauge section, the Cauchy stress tensor, strain tensor, and cyclic frequency. Assuming that the cross section of the specimen in the gage section is constant, the temperature distribution in one-dimensional model with a heat power $\dot{Q}$is [36]:

$$\frac{\partial^2 T}{\partial z^2} - \frac{hP}{Ak}\left(T - T_\infty\right) - \epsilon\beta\frac{P}{Ak}\left(T^4 - T_\infty^4\right) - \frac{\dot{Q}}{k} = \frac{1}{\alpha}\frac{\partial T}{\partial t} \quad (2)$$

where P is the perimeter of the cross section of the sample and α is the thermal diffusivity. Since T is the absolute temperature ($T - T_\infty \ll T_\infty$), it is feasible to linearize the third term by Taylor expansion around T_∞. Thus, Eq. (5) will be reduced to the following partial differential equation (PDE) [36]:

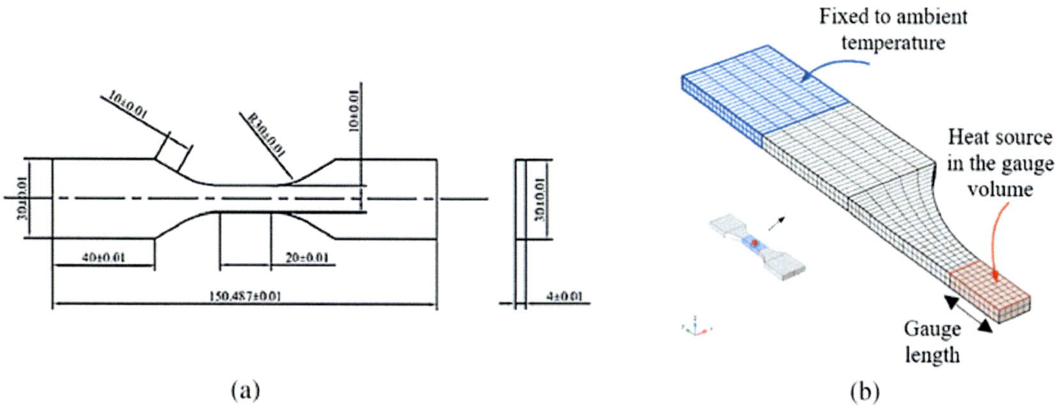

Fig. 1 (**a**) Sample geometry for comparison with an analytical approach [23] and (**b**) FEM modeling

$$\begin{cases} \frac{\partial^2\theta}{\partial z^2} - m^2\theta + \frac{\dot{Q}}{k} = \frac{1}{\alpha}\frac{\partial\theta}{\partial t} \\ \theta(0,t) = \theta(L,t) = 0 \ \ , \theta(z,0) = 0 \end{cases} \tag{3}$$

where $\theta = T - T_\infty$, $m = \sqrt{\frac{(h+4\epsilon\beta T_a^3)P}{Ak}}$ and L is the effective length. Maple programming language was utilized to solve Eq. (3) which is a linear non-homogenous partial differential heat equation. The solutions are as follows:

- For a constant heat, $\dot{Q} = H\left(\frac{W}{m^3}\right)$

$$\theta(z,t) = \frac{1}{km^2\left(e^{2mL}-1\right)}\left\{\left(e^{mz} - e^{m(L+z)} - e^{m(2L-z)} + e^{2mL} + e^{m(L-z)} - 1\right)H + \left(\sum_{n=1}^{\infty}\frac{2\sin\left(\frac{n\pi z}{L}\right)e^{-\frac{\alpha t\left(L^2m^2+\pi^2n^2\right)}{L^2}}HL^2\left((-1)^n-1\right)}{\pi nk\left(L^2m^2+\pi^2n^2\right)}\right)e^{(2mL-1)}m^2k\right\} \tag{4}$$

- For a cyclic heat, $\dot{Q} = H\left(1+\sin(2\pi ft)\right)\left(\frac{W}{m^3}\right)$

$$\theta(z,t) = \sum_{n=1}^{\infty}\frac{1}{kn\pi\left(L^2m^2+\pi^2n^2\right)\left(\left(L^2m^2+\pi^2n^2\right)^2\alpha^2+\omega^2L^4\right)}\left\{2\left(-1+(-1)^n\right)HL^2\sin\left(\frac{n\pi z}{L}\right)\left(\left(\left(L^2m^2+\pi^2n^2\right)^2\alpha^2 - \omega L^2\left(L^2m^2+\pi^2n^2\right)\alpha+\omega^2L\right)e^{-\frac{\alpha t\left(L^2m^2+\pi^2n^2\right)}{L^2}} + \omega L^2\left(L^2m^2+\pi^2n^2\right)\alpha\cos(\omega t) - \left(L^2m^2+\pi^2n^2\right)\alpha^2\sin(\omega t) - \left(L^2m^2+\pi^2n^2\right)^2\alpha^2+\omega^2L^4\right)\right\} \tag{5}$$

where H is the mean value of each heat dissipated power. As shown in Fig. 1a, due to the symmetry of the dog bone sample, only 1/8 of the geometry was modeled with appropriate boundary conditions, shown in Fig. 1b. Hexahedral elements with quadratic Lagrange discretization are used for meshing, and after mesh convergence study, it was meshed with 1260 elements. The temperature at both ends (the domain in blue) was assumed to be fixed to the environmental temperature, T_∞.

For comparison, $H = 10^7 (\text{W/m}^3)$, a thickness of 6 mm, and the C45 steel as the material with parameters listed in Table 1 (all taken from the study by De Finis et al. [23]) were considered. As shown in Fig. 2, the FEM and analytical results are comparable for both constant and cyclic heat source.

In the last second of the simulation, the mean temperatures from numerical and analytical results are 26.164 °C and 26.125 °C, respectively, which give only 0.15% error. In addition, in terms of the temperature amplitude, the error is

Table 1 Input parameters for comparison between the analytical and numerical results; for material C45 [23]

ρ (kg/m^3)	C_p(J/kg ° C)	k(W/m ° C)	α (1/ ° C)	h(W/m^{2}K)	T_∞(°C)	ϵ	f(Hz)
7850	486	42.5	13.3	5	20	0.98	11

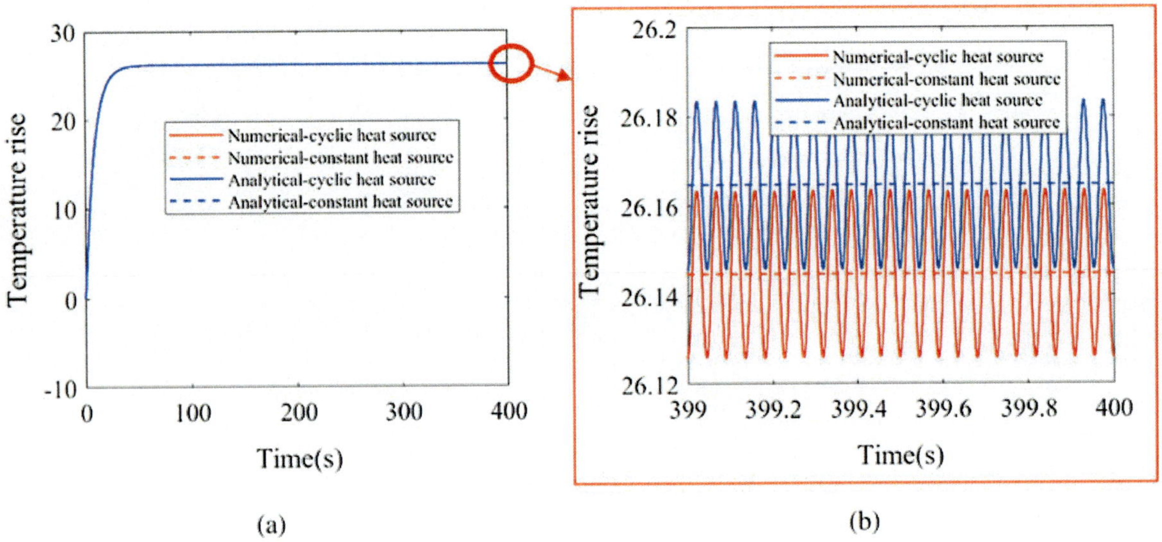

Fig. 2 (**a**) Numerical and analytical results comparison for constant heat source and (**b**) cyclic heat source

about 2.09% with the amplitudes of 0.0191 °C and 0.0195 °C related to the numerical simulation and analytical solution, correspondingly.

Heat Source Identification

In the previous section, the correctness of the numerical model was shown using an analytical 1D model. In this section, the heat source producing the temperature values in terms of second harmonic temperature variations is simulated considering also the effect of geometry. The aim is to investigate differences between the experimentally measured values of SAH and the same values determined from numerical model.

For this purpose, the experimental results were taken from a recent work by De Finis et al. [23] which focused specifically on the possibility of utilizing the SAH of temperature signal as an indicator to predict the heat dissipated energy. As mentioned by Krapez et al. [35], this temperature component completely related to the intrinsic dissipation especially in fully reversed loading. In recent years, researchers have shown their interests in this temperature component and recognized it as an effective index on the fatigue limit prediction [21–23, 33–35, 37].

In general, as mentioned by De Finis et al. [23], a cyclic load causes periodic behaviors for both strain and stress during the fatigue test. Consequently, according to the first law of thermodynamic, in terms of energy rates and ignoring the stored energy rate, the dissipated heat during a cyclic process (under fully reversed loading, $R = -1$) is almost equal to mechanical energy $\dot{E}_d \approx \dot{W}$; therefore, the heat dissipated energy rate, $\dot{E}_d$, is also a periodic function [23]. Imposing DFT (discrete Fourier transform), the second amplitude harmonic of $\dot{E}_d$, $\dot{E}_{d_a}^{2\omega}$, can be evaluated based on the SAH of temperature as the fundamental dissipative temperature component [23]:

$$\dot{E}_{d_a}^{2\omega} = 2f\rho C T_a^{2\omega} \tag{6}$$

where $2f$ shows twice the frequency of the loading and $T_a^{2\omega}$ is the second harmonic amplitude of the temperature associated to E_d and imposing DFT [23].

The experimental value of $\dot{E}_{d_a}^{2\omega}$ for the loading step with stress amplitude of 340 MPa and fully reversed loading condition is 9.59×10^5 (W/m^3) [23]; by using Eq. (6) and the measured value of SAH: $T_{a\,\exp}^{2\omega}$ =0.0114 (°C). Of course, since $T_a^{2\omega}$ was from the surface temperature data, $\dot{E}_{d_a}^{2\omega}$ is only an estimation of its real value [23].

Table 2 Surface $T^{2\omega}_{a_{\text{num}}}$ for different heat source amplitude $\dot{E}^{2\omega}_{d_a}$ and $\dot{W}^{2\omega}_a$ and comparison with the corresponding experimental value

Heat source (W/m^3)		Gauge volume fraction				
		1/15	2/15	5/15	10/15	1
$d = \dot{E}^{2\omega}_{d_a}\ (1 + \sin(2\pi\ (2f)\ t))$	$T^{2\omega}_{a_{\text{num}}}$ (°C)	0.0002	0.0003	0.0007	0.0014	0.0018
	$T^{2\omega}_{a_{\text{exp}}}/T^{2\omega}_{a_{\text{num}}}$	54.2857	33.5294	15.4054	8.2014	6.1957
$d = \dot{W}^{2\omega}_a\ (1 + \sin(2\pi\ (2f)\ t))$	$T^{2\omega}_{a_{\text{num}}}$ (°C)	0.0092	0.0137	0.0315	0.0597	0.0789
	$T^{2\omega}_{a_{\text{exp}}}/T^{2\omega}_{a_{\text{num}}}$	1.2391	0.8321	0.3619	0.1910	0.1445

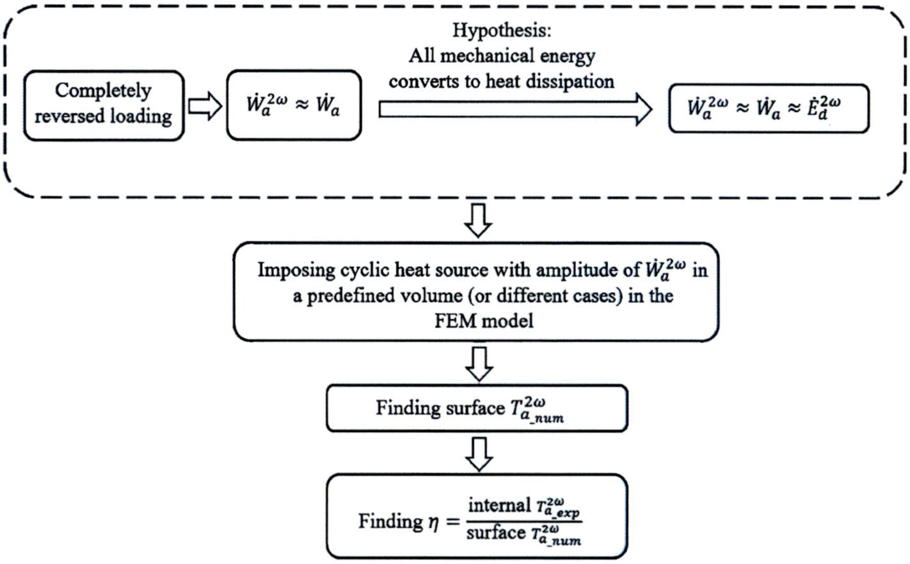

Fig. 3 A schematic representation of heat source identification

Using this value, the geometry, and mechanical properties proposed in Fig. 1 and Table 1, the simulations were performed for four fractions of the gauge volume (the gauge volume is shown in Fig. 1b) in addition to the whole gauge volume because the volume producing heat is unknown. It should be noted that smaller volumes are considered to be the same as the gauge volume, but they are shrunk only in y direction from both ends to provide each fraction. Similar to the experimental procedure in thermography-based technique, the temperature data obtained from the simulation are examined only from the surface of the gauge volume. The input parameters, mesh, and boundary conditions are the same as previous section, and the cyclic heat source was $d = \dot{E}^{2\omega}_{d_a}\ (1 + \sin(2\pi\ (2f)\ t))\ \left(W/m^3\right)$.

As shown in Table 2, comparing experimental $T^{2\omega}_a$ with the numerical one, reveals a big difference even considering the whole of the gauge length producing heat which is not prevalent in the fatigue test. This difference is associated to different causes; one of them can be related to the procedure of measuring the dissipated heat which is based on the acquisition of surface temperature.

Since the experimental value of $\dot{E}^{2\omega}_{d_a}$ was measured from the surface, it cannot be a representative of the real heat source at twice the frequency of the loading; thus it is worth utilizing a much more reliable experimental value: the SAH of the mechanical energy rate, $\dot{W}^{2\omega}_a$, that can be obtained from $\dot{W}$ (strains from extensometer and stresses from loading machine) after applying DFT [23].

It should be noted that in a fully reversed loading, the second harmonic component of the mechanical energy is almost equal to the total mechanical energy [19]. For $R = -1$ and the considered material, it can be also assumed that the mechanical energy is totally converted to heat, so under these conditions $\dot{W}^{2\omega}_a$ can be used as the parameter representing the total heat dissipation. For clarification, the procedure is schematically presented in Fig. 3.

By considering a cyclic heat source in the form of $d = \dot{W}^{2\omega}_a\ (1 + \sin(2\pi\ (2f)\ t))\ \left(\text{W/m}^3\right)$, as shown in Table 2 and $\dot{W}^{2\omega}_a = 4.11 \times 10^7\ \left(\text{W/m}^3\right)$ [23], the surface $T^{2\omega}_{a_{\text{num}}}$ approaches to $T^{2\omega}_a$ exp when the heat volume is between 1/15 and 2/15 of the gauge volume. This portion of volume seems to show that the plastic deformation occupied not the whole but a portion of the gauge volume as it can be expected in the fatigue failure. Thus, at this stage, the heat volume that can produce acceptable $T^{2\omega}_a$on the surface of the gauge volume is found, numerically. Of course, more investigations to explain physically this result are required.

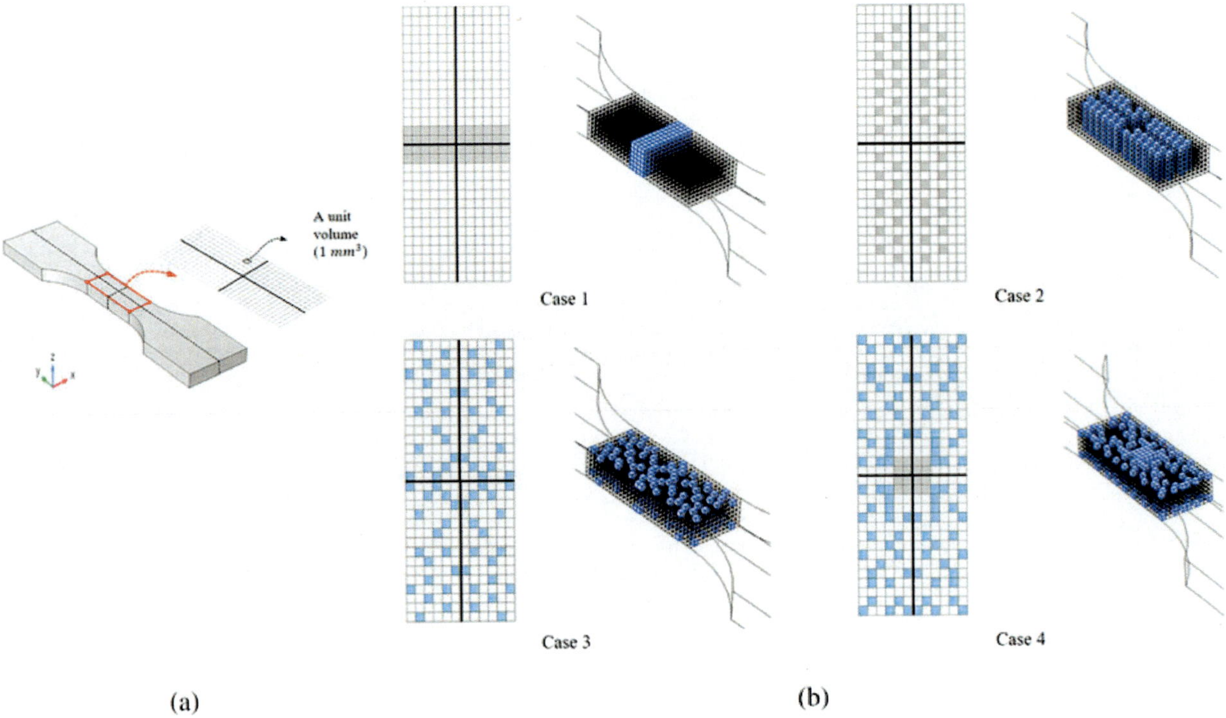

Fig. 4 (**a**) Grid on the surface of the gauge length; (**b**) representation of different heat source geometries

Table 3 Geometry of each volume(s) in each case

Case #	Number of separated volumes	Length (mm) of volume edge in		
		x-direction	y-direction	z-direction
1	1	12	4	6
2	48	1	1	6
3	144	1	1	2
4	1	4	4	6
	192	1	1	1

Furthermore, it is of great importance investigating the influence of the geometry of the heat volume on surface $T_a^{2\omega}$ since considering a single integrated volume, for example, 2/15 of the gauge volume as presented in Fig. 4b and Case 1, is almost impossible during the fatigue test, since there are lots of regions (i.e., the defects existing both inside and on the surface of the material) which can produce heat when deformed plastically.

For this purpose, as can be seen in Fig. 4b, the geometry of the heat source was changed for three more cases to check what happens to the temperature amplitude. A schematic representation is proposed to introduce different cases. As shown in Fig. 4a, the red border, the surface of the gauge volume, contains a set of squares (or elements) representing the top surface of a cube with the volume of 1 mm^3 where a heat source can be located. The thick black horizontal and vertical lines on the grid show the symmetric planes of the dog bone sample. The considered cases are presented with two colors, and each color shows how the heat source is spread through the thickness. The gray color is for a volume with boundary from the top to bottom of the gauge volume, while the blue color is related to volumes spread only on the top and bottom surfaces of the gauge volume. It should be noted that all cases are assumed to have a fixed volume of 2/15 gauge volume as this fraction provides acceptable surface $T_a^{2\omega}$ (please see Table 2). In addition, the geometry of the volume(s) in each case is listed with details in Table 3. In Case 4, a more real form of a heat volume was modeled. It is assumed that there are some small volumes on the surface (they can be a representative of the damage accumulation sites and possible crack triggers) with a relatively big one in the center of the gauge length (it shows damage accumulation which is highly possible to be presented with a fully reversed load).

From Table 4, the surface $T_{a_{\text{num}}}^{2\omega}$ is significantly affected by the heat source geometry in which it decreases by 10.95% for Case 2 and increases, respectively, by 37.96% and 106.57% for Case 3 and 4, if compared with Case 1.

Table 4 Numerical surface $T_a^{2\omega}$ with $d = \dot{W}_a^{2\omega}\ (1 + \sin(2\pi\ (2f)\ t))\ \left(W/m^3\right)$ for different heat volume geometries

	Surface $T_{a_{\text{num}}}^{2\omega}$	$\eta = \frac{\text{internal } T_{a_{\text{exp}}}^{2\omega}}{\text{surface } T_{a_{\text{num}}}^{2\omega}}$
Case 1	0.0137	35.69
Case 2	0.0122 (-10.95%)	40.08
Case 3	0.0189 (+37.96%)	25.87
Case 4	0.0283 (+106.57%)	17.28

In addition, for the heat amplitude equal to $\dot{W}_a^{2\omega}$, a comparison between numerical surface $T_a^{2\omega}$ where temperature is measured and the related value inside the gauge volume, called internal$T_{a_{\text{exp}}}^{2\omega}$, is presented. As assumed previously and shown in Fig. 3, with the total heat dissipation almost equal to $\dot{W}_a^{2\omega}$, the internal$T_{a_{\text{exp}}}^{2\omega}$ can be calculated to be equal to 0.489 directly by replacing $\dot{W}_a^{2\omega}$ instead of $\dot{E}_{d_a}^{2\omega}$ in Eq. (6). For this purpose, the ratio of internal $T_{a_{\text{exp}}}^{2\omega}$ to the surface$T_{a_{\text{num}}}^{2\omega}$ (η) is calculated and compared with the experimental one in the work by De Finis [23], which was almost equal to 20. Note that they found a linear relation between $\dot{W}_a^{2\omega}$ and $\dot{E}_{d_a}^{2\omega}$ [23] which is the same as the relation between $T_{a_{\text{exp}}}^{2\omega}$ and $T_a^{2\omega}$ as the latter two can be obtained by dividing the related energy components by $2f\rho C_p$ (see Eq. (6)).

It can be seen from the last lines (Case studies 3 and 4) of Table 4 that η is relatively close to the above experimental ratio of 20. It shows the volume in the form of Case 4 can be a possible geometry for the real heat source in the fatigue test done by De Finis [23]. As a conclusion, more trial heat volume geometries can also be examined numerically and checked in the same way for the considered four cases to understand better the volumes producing heat, in future works.

Conclusion

In this work, a numerical model has been proposed to estimate the heat source in terms of the amplitude and geometry. It was compared with an analytical solution of a one-dimensional heat equation to be ensured if the model works precisely. Comparing with the analytical solutions for constant and cyclic heat sources, the errors were below 5% which is evident for the accuracy of the model. The results from modelling four possible heat volumes show that the SAH of temperature component, as the fundamental dissipative temperature component, is significantly dependent on both the size and shape of considered volumes.

In addition, comparing the ratio of the numerical and experimental SAHs of surface temperature to the experimental one inside the gauge volume is shown to provide an insight into the real heat volume identification.

This study can be regarded as preliminary work aimed to investigate the heat volumes created during the fatigue test and determine the relationship between the actual heat sources and the superficial temperatures obtained via thermographic measurements. Future works will investigate more heat volumes and will study other geometries of the samples and other thermal components.

References

1. Schijve, J.: Fatigue of Structures and Materials. Springer, Dordrecht (2009)
2. Bjørheim, F., Siriwardane, S.C., Pavlou, D.: A review of fatigue damage detection and measurement techniques. Int. J. Fatigue. **154**, 106556 (2022)
3. Subra, S.: Fatigue of Materials. Cambridge Solid State Science Series (1992)
4. Mughrabi, H.: Cyclic slip irreversibilities and the evolution of fatigued. Metall. Mater. Trans. A. **40**, 1257–1279 (2009)
5. Mareau, C., Favier, V., Weber, B., Galtier, A., Berveiller, M.: Micromechanical modeling of the interactions between the microstructure and the dissipative deformation mechanisms in steels under cyclic loading. Int. J. Plast. **32–33**, 106–120 (2012)
6. Vollmer, M., Möllmann, K.-P.: Infrared Thermal Imaging: Fundamentals, Research and Applications. Wiley (2011)
7. Maldague, X.: Theory and Practice of Infrared Technology for Nondestructive Testing. Wiley (2001)
8. Geraci, A., La Rosa, G., Risitano, A.: L'infrarosso termico nelle applicazioni meccaniche. ATA Ingegneria Automotoristica Catania, Italy (1983)
9. Luong, M.: Infrared thermographic scanning of fatigue in metals. Nucl. Eng. Des. **158**, 363–376 (1995)
10. Enke, N.F., Sandor, B.I.: Cyclic plasticity analysis by differential infrared thermography. In: Proceedings of the VII International Congress on Experimental Mechanics, pp. 830–835 (1988)
11. Kaleta, J., Blotny, R., Helmuth, H.H.: The accumulated internal energy in the fatigue strength region. In: Proceedings of the 7th International Conference on Fracture, pp. 1195–1202 (1989)
12. Luong, M.: Fatigue limit evaluation of metals using an infrared thermographic technique. Mech. Mater. **28**, 155–163 (1998)

13. Fargione, G., Geraci, A., Rosa, G.L., Risitano, A.: Rapid determination of the fatigue curve by the thermographic method. Int. J. Fatigue. **24**, 11–19 (2002)
14. Curà, F., Curti, G., Sesana, R.: A new iteration method for the thermographic determination of fatigue limit in steels. Int. J. Fatigue. **27**, 453–459 (2005)
15. Boulanger, T., Chrysochoos, A., Mabru, C., Galtier, A.: Calorimetric analysis of dissipative and thermoelastic effects associated with the fatigue behavior of steels. Int. J. Fatigue. **26**, 221–229 (2004)
16. Maquin, F., Pierron, F.: Heat dissipation measurements in low stress cyclic loading of metallic materials: from internal friction to micro-plasticity. Mech. Mater. **41**, 928–942 (2009)
17. Naderi, M., Khonsari, M.M.: An experimental approach to low-cycle fatigue damage based on thermodynamic entropy. Int. J. Solids Struct. **47**, 875–880 (2010)
18. Wang, X., Crupi, V., Jiang, C., Guglielmino, E.: Quantitative Thermographic Methodology for fatigue life assessment in a multiscale energy dissipation framework. Int. J. Fatigue. **81**, 249–256 (2015)
19. De Finis, R., Palumbo, D., Galietti, U.: A multianalysis thermography-based approach for fatigue and damage investigations of ASTM A182 F6NM steel at two stress ratios. Fatigue Fract. Eng. Mater. Struct. **42**, 267–283 (2019)
20. Lipski, A.: Change of specimen temperature during the monotonic tensile test and correlation between the yield strength and thermoelasto-plastic limit stress on the example of aluminum alloys. Materials. **14**(1), 13 (2021)
21. De Finis, R., Palumbo, D., Pirinu, A., Saponaro, A., Panella, F.W., Nobile, R., Galietti, U.: Fatigue behaviour assessment of C45 steel by means of energy-based methods. IOP Conf. Ser. Mater. Sci. Eng. **1038** (2021)
22. Meneghetti, G., Ricotta, M.: Estimating the intrinsic dissipation using the second harmonic of the temperature signal in tension-compression fatigue, Part II: experiments. Fatigue Fract. Eng. Mater. Struct. **44**, 2153–2167 (2021)
23. De Finis, R., Palumbo, D., Galietti, U.: On the relationship between mechanical energy rate and heat dissipated rate during fatigue for a C45 steel depending on stress ratio. Fatigue Fract. Eng. Mater. Struct. **44**, 2781–2799 (2021)
24. Fan, J., Guo, X., Zhao, X.: An energetic method to evaluate the macro and micro high-cycle fatigue behavior of the aluminum alloy. Proc. Inst. Mech. Eng. Part C J. Mech. Eng. Sci. **232**, 1456–1469 (2018)
25. Guo, Q., Guo, X., Fan, J., Syed, R., Wu, C.: An energy method for rapid evaluation of high-cycle fatigue parameters based on intrinsic dissipation. Int. J. Fatigue. **80**, 136–144 (2015)
26. Wang, X., Crupi, V., Jiang, C., Feng, E., Guglielmino, E., Wang, C.S.: Energy-based approach for fatigue life prediction of pure copper. Int. J. Fatigue. **104**, 243–250 (2017)
27. Guo, Q., Zaïri, F., Yang, W.: Evaluation of intrinsic dissipation based on self-heating effect in high-cycle metal fatigue. Int. J. Fatigue. **139**, 105653 (2020)
28. Mehdizadeh, M., Khonsari, M.M.: On the role of internal friction in low-and high-cycle fatigue. Int. J. Fatigue. **114**, 159–166 (2018)
29. Guo, Q., Guo, X.: Research on high-cycle fatigue behavior of FV520B stainless steel based on intrinsic dissipation. Mater. Des. **90**, 248–255 (2016)
30. Zhao, A., Xie, J., Zhao, Y., Liu, C., Zhu, J., Qian, G., Wang, S., Hong, Y.: Fatigue limit evaluation via infrared thermography for a high strength steel with two strength levels. Eng. Fract. Mech. **268**, 108460 (2022)
31. Colombo, C., Sansone, M., Patriarca, L., Vergani, L.: Rapid estimation of fatigue limit for C45 steel by thermography and digital image correlation. J. Strain Anal. Eng. Des. **56**, 478–491 (2021)
32. Meneghetti, G.: Analysis of the fatigue strength of a stainless steel based on the energy dissipation. Int. J. Fatigue. **29**, 81–94 (2007)
33. Shiozawa, D., Inagawa, T., Washio, T., Sakagami, T.: Fatigue limit estimation of stainless steels with new dissipated energy data analysis. Proc. Struct. Integr. **2**, 2091–2096 (2016)
34. Krapez, J.-C., Pacou, D., Bertin, C.: Application of lock-in thermography to rapid evaluation of fatigue limit in metals. In: Workshop on Advanced Infrared Technology and Applications (1999)
35. Krapez, J.-C., Pacou, D., Gardette, G.: Lock-in thermography and fatigue limit of metals. Quant. InfraRed Thermogr. (2000)
36. Jang, J., Khonsari, M.M.: On the evaluation of fracture fatigue entropy. Theor. Appl. Fract. Mech. **96**, 351–361 (2018)
37. Meneghetti, G., Ricotta, M.: Estimating the intrinsic dissipation using the second harmonic of the temperature signal in tension-compression fatigue: Part I. Theory. Fatigue Fract. Eng. Mater. Struct. **44**, 2168–2185 (2021)

A Full Experimental Characterization of Aged Nitrile Butadiene Rubbers (NBR)

Adrien Redon, Jean-Benoît Le Cam, Eric Robin, Mathieu Miroir, and Jean-Charles Fralin

Abstract Part of the sealing function in the pneumatic system of the French high-speed train (TGV) is provided by elastomeric components, such as nitrile butadiene rubber (NBR) O-rings. Like most elastomers, NBRs are sensitive to thermo-oxidative aging. In the current application, aging results in a strong hardening of the O-rings, which alters the sealing function and impacts the lifetime of the mechanical pneumatic system. Therefore, understanding the origin of this aging, studying its effects on the mechanical properties, and modeling it are of a paramount importance in optimizing time in maintenance operations of the TGV.

In a previous study, accelerated aging tests have been carried out with different NBR formulations at different times and temperatures. These tests enabled us to reproduce ex situ the hardening observed on the O-rings. Aging mechanisms have been identified and related to the aging conditions of the NBR formulations through different characterization techniques: infrared spectroscopy, swelling tests, X-ray fluorescence spectrometry, differential scanning calorimetry thermogravimetric analysis, and micro-hardness.

To consider the effects of the mechanical loading on the aging process and the mechanical behavior of tested NBRs, mechanical spectroscopy (DMA) analyses have been carried out. The results obtained highlight how mechanical loadings impact the aging of these NBRs and quantify the effects of aging on their viscoelasticity. From the collected data, a hyperelastic model has finally been identified to simulate the behavior of O-rings by the finite element method. These first results lay the foundation of a selection methodology for O-rings used in the pneumatic system of the TGV.

Keywords Elastomer · Aging · DMA · Lifetime prediction

Introduction

Nitrile butadiene rubbers (NBRs) are extensively used in sealing applications with O-rings, due to excellent chemical resistance to oils and greases, good mechanical properties, as well as moderate production costs. Nevertheless, several aging processes alter their mechanical behavior, which decreases the lifetime of the parts made with them.

The present study focuses on commercial NBR O-rings, used for the sealing function in the pneumatic system of the French high-speed train. During their lifetime, these O-rings are exposed to prolonged contact with oxygen and to temperatures ranging between −30 °C and 50 °C. These factors result in a significant hardening, as well as permanent deformation, which leads to a loss of sealability, since the distorted O-ring loses its elasticity. This forces Société Nationale

A. Redon (✉)
Univ Rennes, CNRS, IPR (Institut de Physique de Rennes) – UMR 6251, Rennes, France

SNCF Voyageurs, Ingénierie du Matériel (CIM OMF), Rennes Cedex 9, France
e-mail: adrien.redon@univ-rennes1.fr

J.-B. Le Cam · E. Robin · M. Miroir
Univ Rennes, CNRS, IPR (Institut de Physique de Rennes) – UMR 6251, Rennes, France
e-mail: adrien.redon@univ-rennes1.fr; jean-benoit.lecam@univ-rennes1.fr; eric.robin@univ-rennes1.fr; mathieu.miroir@univ-rennes1.fr

J.-C. Fralin
SNCF Voyageurs, Ingénierie du Matériel (CIM OMF), Rennes Cedex 9, France
e-mail: jean-charles.fralin@sncf.fr

C. Franck et al. (eds.), *Challenges in Mechanics of Biological Systems and Materials, Thermomechanics and Infrared Imaging, Time Dependent Materials and Residual Stress, Volume 2*, Conference Proceedings of the Society for Experimental Mechanics Series,
https://doi.org/10.1007/978-3-031-50470-9_13

des Chemin de fer Français (SNCF, France's national railway company) to multiply the number of maintenance operations and to replace only the hardened O-rings, whereas all the other components of the pneumatic system are still functional.

Prolonged exposure to heat and oxygen is a known aggravating factor in aging of NBRs and more generally of elastomers. In a previous study [1], the effect of thermo-oxidative aging on eight different filled nitrile rubbers was highlighted with accelerated aging tests. The formation of oxidative functionalities and consumption of unsaturations (double bonds) in the butadiene part of the copolymer during aging led to a degradation of the mechanical properties (i.e., micro-hardness). Also, swelling test showed that extra crosslinks form in the elastomers, increasing their hardness and reducing the mobility of the macromolecular chains, making the elastomers hard and brittle. A comparison with aged O-rings from trains was made and illustrated how these O-rings lose their sealing function: they are "frozen" in a deformed state due to both mechanical loadings and aging (loss of elasticity), which leads to leaks in the systems they equip.

As abovementioned, mechanical loadings are key contributors to differences between accelerated thermal aging and actual aging behavior in the industrial application. As such, they need to be investigated thoroughly. The main interest of the present study is to compare different NBR formulations, submitted to the same aging conditions, before and after aging. The results aim at reproducing accurately ex situ the aging observed on trains after several years. The comparison enables us to identify and to correlate aging mechanisms to variations in viscoelastic properties. Establishing this relationship is needed to propose a suitable hyperelastic model and to perform a reliable simulation of the behavior of O-rings by the finite element method.

In the first part of this article, the experimental protocol is given, as well as background information of the different tests and methods used. The second part focuses on the results obtained, before and after aging, at different aging times and temperatures. This part lays the foundation for a future selection methodology of O-rings used in the pneumatic system of the French high-speed train.

Experimental Protocol

Materials and Methods

NBRs are butadiene (C_4H_6) and acrylonitrile (C_2H_3CN) copolymers. The butadiene monomer polymerizes into three distinct isomers, namely, *trans-1,4*, *cis-1,4*, and *vinyl-1,2*. The acrylonitrile (ACN) monomer is statistically inserted within the macromolecular chain, and its content confers to NBRs their main properties, such as oil resistance. It also influences the glass transition temperature (T_g). ACN content can be adjusted in a range generally going from 20% to 45%. For instance, for low-temperature applications, ACN content is found around 20% to have low values of T_g, i.e., less than −40 °C.

In the present study, nine types of commercial NBRs are considered, provided by four different suppliers, denoted A, B, C, and D. They are classified according to their glass transition temperature and their original hardness and are presented in Table 1.

Compression set type A disks, with a diameter of 29.0 mm and a thickness of 12.5 mm (NF ISO 815-1), are used for the different tests. Each material is prepared from the same material batch by the suppliers. Temperature and frequency sweep tests are performed on all the unaged rubber samples. Three types of rubbers (A1-4070, B1-4070, and D1-4070) that better suit the industrial application were chosen for compression set (CS) aging and further DMA testing.

Table 1 NBRs considered in this study

Specimen code	T_g (°C)	Hardness (IRHD)
A1-4070	−54	70
A2-4050	−56	50
A3-2070	−28	70
B1-4070	−48	70
B2-4050	−55	50
C1-4070	−45	70
C2-4050	−45	50
D1-4070	−50	70
D2-4050	−50	50

Thermo-oxidative Aging and Compression Set (CS)

Accelerated aging tests were performed at 100 °C in a Memmert (ULP 500) air-circulating oven, for 72 h, 168 h, 336 h, 672 h, 1176 h, and 2016 h. Compression set rigs are used to compress the rubber specimens at 15%, before putting the specimens in the oven. After aging, the compression set rig is cooled at room temperature for 30 min, before taking out the compressed specimens. The height of the specimens is measured after an additional 30 min, and CS values are obtained according to the following formula:

CS (%) $= \frac{h_0 - h_1}{h_0 - h_s} \times 100$ with h_0 the original height of the specimen, h_s the spacer height, and h_1 the height after testing. Temperature measurements are made by a 4-wire Pt100 sensor, with a thermal resolution of 1 K. A fan is used for forced air circulation. For each aging time and temperature, three specimens per material were tested. Future investigations are planned at 70 °C and 85 °C to study the effect of temperature on the aging tests and will be presented during the conference.

DMA Tests

The viscoelastic behavior of a material can be interpreted in the time domain, as a relaxation modulus $E(t)$, or in the frequency domain, by the dynamic modulus, denoted $E^*(\omega)$, with ω the angular frequency. $E^*(\omega)$ can be experimentally characterized by dynamical mechanical analysis (DMA) tests. In DMA experiments, a sinusoidal force or displacement is applied to the test specimens, and the resulting viscoelastic response of the studied material is measured. If the strain imposed takes the form of $\varepsilon(t) = \varepsilon_d \sin(\omega t)$, with ε_d the dynamic strain amplitude, the resulting stress will follow the equation $\sigma(t) = \sigma_d \sin(\omega t + \delta)$ with δ the angular shift (see Fig. 1). The dynamic modulus can be decomposed as a storage modulus (E') and a loss modulus (E''), such that $E^*(\omega, \varepsilon_d) = \frac{\sigma_d}{\varepsilon_d}(\cos\delta + i \sin\delta) = E'(\omega, \varepsilon_d) + i E''(\omega, \varepsilon_d)$ [2]. E' and E'' are calculated from the material response to the sinusoidal wave and represent, respectively, the ability of said material to return or store energy (E') and its ability to lose it (E''). A loss (or damping) factor is conventionally defined as $\tan\delta = E''/E'$.

Therefore, the viscoelastic properties of the different filled nitrile rubbers are measured using mechanical spectroscopy (DMA) analyses, using the compression set disks aged from compression set tests. For all specimens, temperature and frequency-temperature sweep (FTS) tests are performed. Temperature sweep tests give us basic dynamic mechanical properties, such as E', E'', and $\tan\delta$, as a function of temperature, ranging from −40 °C to 100 °C, with a heating rate of 2 °C/min. The test frequency is fixed at 1 Hz, and the dynamic strain rate is set at 0.001%. To maintain the disk under compression loading, a static force of −100 N is applied.

For FTS tests, master curves of E', E'', and $\tan\delta$ are obtained by using the time-temperature superposition principle [3] for different aging times. This characterization is performed for frequencies ranging between 1 and 10 Hz, with dynamic strain rate set at 0.001% and temperature isotherms from −20 °C to 80 °C, with a heating rate of 2 °C/min. For each isotherm, a 10-min wait is imposed to make sure measurements are taken at the right temperature. A static force of −50 N is applied.

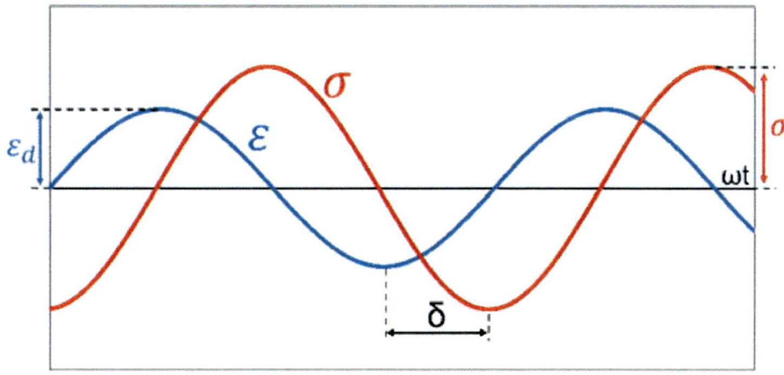

Fig. 1 DMA principle, stress, strain and phase shift behavior of viscoelastic material

Results and Discussions

Unaged Specimens

The variations of basic dynamic mechanical properties, such as storage modulus E', loss modulus E'', and loss factor tanδ,provide fundamental knowledge in understanding the behavior of filled rubbers. For unaged specimens, temperature sweeps and frequency sweeps are performed and presented in Figs. 2 and 3.

From Fig. 2 and as expected, the different 70 IRHD specimens present similar viscoelastic behavior, since the specimens are unaged and fit the same technical specifications. Nevertheless, the suppliers are different; therefore the aging behavior should differ with aging time. In Fig. 3, experimental characterization is done between 1 and 10 Hz, for different temperatures. The experimental master curve is calculated using time-temperature superposition principle and Williams Landel and Ferry equation (WLF) [3] for elastomers: $\log a_T = \frac{C_1(T-T_{\text{ref}})}{C_2+(T-T_{\text{ref}})}$, where a_T is the shift factor, T is the temperature, T_{ref} is a reference temperature chosen to construct the master curve, and C_1 and C_2 are constants. The master curve can also be fitted using a polynomial approach, which is shown in Fig. 3. Unaged characterization of the specimens is important to further compare the specimens after aging and to investigate how aging affects the viscoelastic and dynamic properties of studied NBRs.

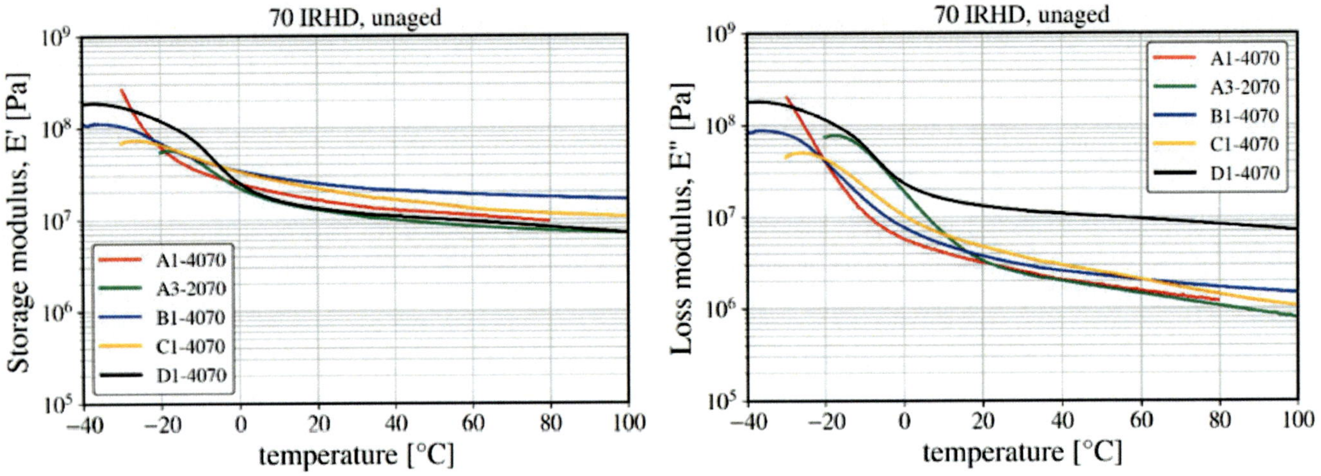

Fig. 2 Example of temperature sweep tests, with E' and E'' values for unaged 70 IRHD specimens

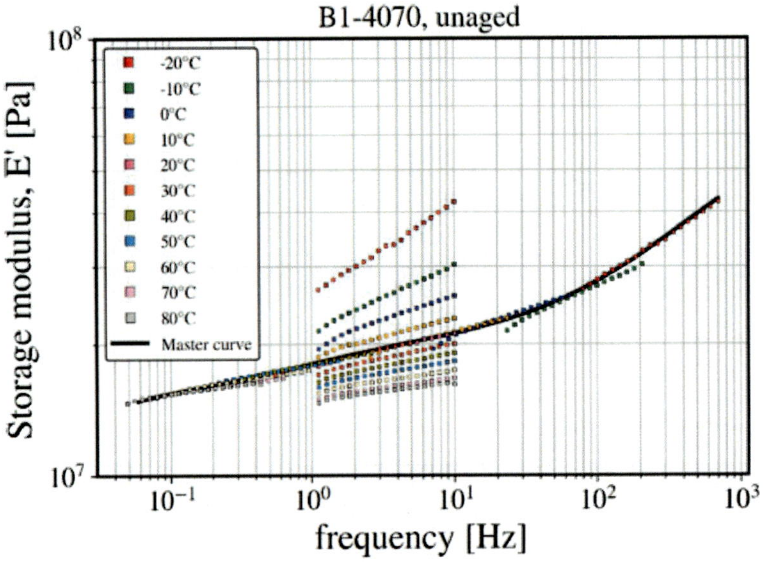

Fig. 3 Example of an FTS test, with E' values for B1-4070. The master curve is calculated using time-temperature superposition principle and fitted by the polynomial approach

CS Measurements After Aging

From the nine types of commercial NBRs considered in the study, three of them (A1-4070, B1-4070, and D1-4070) that better suit the industrial application are selected. Accelerated aging tests at 100 °C are performed under compression of 15%, using compression set rigs. After aging, the compression set (CS) is measured according to the formula presented in the experimental protocol.

The final variation of CS with aging time will be provided during the conference since measurements are still in progress. First results show that CS increases [4, 5] with aging time and h_1 value decreases. This value represents the thickness of the tested specimen after elastic recovery. It shows, as expected, that the elastic recovery ability of tested specimens decreases and thus its viscoelastic properties.

A link with previous results on the deterioration of NBRs during thermo-oxidative aging [1] can further explain this phenomenon. The deterioration of viscoelastic properties in the studied NBRs is a consequence of the changes of molecular structure and crosslink density caused by thermo-oxidative aging [4]. Relationships between CS, aging time, crosslink densities, and aging temperatures will be investigated.

DMA Measurements After Aging

The variation of dynamic properties with aging time under different aging temperatures will be provided during the conference since measurements are still in progress. E' and E'' are expected to increase with aging time and tanδ to decrease with aging time [6, 7], illustrating the change in viscoelastic properties of filled nitrile rubbers and more particularly a progressive increase in the crosslink density. Higher temperatures should also accelerate the aging process, resulting in higher values of E' and E'' and lower tanδ values. A comparison with previous micro-hardness values, crosslink density [1], and CS values after aging will be done and will provide valuable insights in fully characterizing the aging phenomenon and its consequences on nitrile rubbers.

Conclusion

The effect of mechanical loadings on the viscoelastic properties of different filled nitrile rubbers is addressed in this study. Both CS tests and DMA test show how the different specimens' properties are affected by aging time, aging temperature, and stress. A link with a previous study on thermo-oxidative aging, such as modification of the macromolecular structure and increase in crosslink density, is currently being investigated, regarding viscoelastic properties after aging.

Furthermore, a suitable hyperelastic model, computed from DMA data and derived from a generalized Maxwell model in the discrete case of Prony series, is being addressed [8, 9]. Evolution of Prony parameters during aging is currently being investigated. Establishing relationships between all the abovementioned factors is detrimental to reliably predict the lifetime and behavior of O-rings used in the pneumatic system of the French high-speed train.

Acknowledgments The authors gratefully acknowledge SNCF Voyageurs for their financial support and SNCF's Ingénierie du Matériel (CIM Rennes and AEF) for their technical assistance.

References

1. Redon, A., Miroir, M., Robin, E., Le Cam, J.B., Fralin, J.C.: Aging of NBR: Application to O-rings used in the braking system of the French high-speed train. In: Constitutive Models for Rubber XII, pp. 420–426. CRC Press (2022)
2. Jalocha, D.: Payne effect: a constitutive model based on a dynamic strain amplitude dependent spectrum of relaxation time. Mech. Mater. **148**, 103526 (2020)
3. Williams, M.L., Landel, R.F., Ferry, J.D.: The temperature dependence of relaxation mechanisms in amorphous polymers and other glass-forming liquids. J. Am. Chem. Soc. **77**(14), 3701–3707 (1955)
4. Han, R., Wu, Y., Quan, X., Niu, K.: Effects of crosslinking densities on mechanical properties of nitrile rubber composites in thermal oxidative aging environment. J. Appl. Polym. Sci. **137**(36), 49076 (2020)

5. Qian, Y.H., Xiao, H.Z., Nie, M.H., Zhao, Y.H., Luo, Y.B., Gong, S.L.: Lifetime prediction and aging behaviors of nitrile butadiene rubber under operating environment of transformer. J. Electr. Eng. Technol. **13**(2), 918–927 (2018)
6. Wei, Y.T., Nasdala, L., Rothert, H., Xie, Z.: Experimental investigations on the dynamic mechanical properties of aged rubbers. Polym. Test. **23**(4), 447–453 (2004)
7. Garnier, P., Le Cam, J.B., Grediac, M.: The influence of cyclic loading conditions on the viscoelastic properties of filled rubber. Mech. Mater. **56**, 84–94 (2013)
8. Jalocha, D., Constantinescu, A., Neviere, R.: Revisiting the identification of generalized Maxwell models from experimental results. Int. J. Solids Struct. **67**, 169–181 (2015)
9. Tapia-Romero, M.A., Dehonor-Gómez, M., Lugo-Uribe, L.E.: Prony series calculation for viscoelastic behavior modeling of structural adhesives from DMA data. Ingeniería investigación y tecnología. **21**(2) (2020)

Characterization of the Effect of Cold Joints and Functional Grading in Stepwise Constructed Tensile Dog Bones with Embedded Digital Image Correlation

Tomislav Kosta, Claron J. Ridge, Marcel M. Hatter, and Jesus O. Mares

Abstract We present an extension of our previously published work in which we utilized an in situ/embedded Digital Image Correlation (DIC) technique to evaluate the effect of aging on the interface debonding strength in an idealized particulate composite. The idealized composite consists of a single ~650 μm glass bead inclusion at the center of a Sylgard 184 dog bone tensile sample along with an embedded DIC speckle pattern at the mid-plane. The speckle pattern enables the measurement of the strain field in the region surrounding the embedded glass bead while the sample undergoes a tensile load to failure. The measured strain field can then be used in conjunction with the global load measurement to characterize the local stress required to induce a failure at the interface. In previous works we have demonstrated how this approach can be used to characterize interface debonding. Construction of these tensile samples is done in a stepwise fashion with partial cure cycles in order to achieve a reliable in situ/embedded DIC speckle pattern. This stepwise construction introduces two uncertainties into the experiment, namely, cold joints between the casting layers and functional grading of the stiffness properties due to each layer experiencing different curing histories. In this work we present an evaluation of the effect of the cold joints and functional grading by performing tension tests to measure the global stiffness of dog bones of various sample configurations spanning the number of cold joints and the cure cycle history. By comparing the measured stiffness of each sample configuration, the effect of the cold joints and functional grading are determined.

Keywords DIC · Tension testing · Sylgard · Cold joints · Functional grading

Introduction

The bulk mechanical properties of polymer matrix composites depend highly upon the mechanical properties of the individual components as well as the interfaces between those components. This is especially true under tensile loading where weak interface adhesion can lead to interface debonding which ultimately reduces the integrity and load carrying capability of the material. In particulate composites the total interface area can be significant, especially for high volume fraction filled materials; therefore interface adhesion and strength can have a strong effect on tensile strength properties. Due to the stochastic nature of failure as well as the technical challenges involved in measuring interface debonding in particulate composites, it is difficult to achieve an accurate statistical representation of the debonding process.

Several standard methods exist to measure the adhesion of a simple binary material system; however, these tests are typically conducted to provide qualitative comparisons to indicate increased or decreased adhesion relative to a baseline. Tests such as the peel, [1], pull [2], and blister test [3] provide a means of quantifying the debonding process; however, these tests are not well suited for particulate composite materials due to sample geometry and size requirements. While it is possible to fabricate samples which would enable use of these tests, the data would not be representative due to differences in properties such as surface morphology, etc.

T. Kosta (✉) · C. J. Ridge · J. O. Mares
United States Air Force Research Laboratory, Munitions Directorate, Eglin AFB, FL, USA
e-mail: tomislav.kosta@us.af.mil; claron.ridge.1@us.af.mil; jesus.mares.2@us.af.mil

M. M. Hatter
University of Dayton Research Institute, Dayton, OH, USA
e-mail: Marcel.hatter.ctr@us.af.mil

C. Franck et al. (eds.), *Challenges in Mechanics of Biological Systems and Materials, Thermomechanics and Infrared Imaging, Time Dependent Materials and Residual Stress, Volume 2*, Conference Proceedings of the Society for Experimental Mechanics Series,
https://doi.org/10.1007/978-3-031-50470-9_14

Various approaches have been published which provide a more direct means of characterizing the interface debonding behavior and are more well suited for particulate composites. For example, Tan et al. [4] have recently integrated Digital Image Correlation (DIC) into a mode-I fracture test of a composite with a high volume fraction of particulate filler and an elastomeric matrix material to provide targeted data for calibration of a cohesive zone element (CZE) model. These experiments offer a measurement of interface debonding behavior; however, the statistical distribution of the properties of individual interfaces is partially lost due to the integrated nature of the measurement.

Alternatively, experiments such as those by Gent and Park [5] and proposed by Lauke [6, 7] explicitly consider the debonding of a single filler particle within a matrix material. The work by Gent and Park [5] examines the debonding of a single glass sphere embedded in an optically transparent matrix material. Under a tensile load, the onset of interface debonding was observed, and the local stress state was analytically determined using the load frame force measurement and sample geometry configuration. A critical debonding stress was noted to serve as an indicator of the strength of adhesion between the filler particle and the matrix material.

In previous work [8], we presented an extension of the technique proposed by Gent and Park [5], where we incorporated embedded DIC to better quantify the interface debonding event. Our approach utilizes an embedded DIC speckle pattern to quantify the strain field surrounding the mid-plane of a glass sphere embedded at the center of an optically transparent polymer material. We are able to quantitatively capture the strain field during a tensile loading event from the initial unloaded state up to and through the debonding event. Our approach is briefly summarized in the following section; however, the reader is referred to [8] for a more complete discussion.

In this work, we investigate various aspects of the sample fabrication procedure to determine the effect, if any, on the interpretation of the acquired data. Our approach necessitates a stepwise fabrication of the tensile samples which involves multiple casting and curing steps in order to ensure proper placement of the embedded glass sphere and DIC speckle pattern. An artifact of having multiple cast and cure steps is a functional grading of the elastic properties through the thickness of the tensile samples, as well as cold joints between the various casting layers.

Experimental Method

Summary of Tensile Test Technique

We briefly summarize our technique to characterize interface debonding using an embedded DIC approach here; however, the reader is referred to [8] for a more complete discussion.

Tensile test samples, or "dog bones," are constructed using a multiple-step process to create an embedded glass bead within an elastomeric material with an embedded field of tracking particles at the mid-plane. In order to achieve optical access to the glass bead of interest, a transparent matrix material of polydimethylsiloxane (PDMS) is used. The tensile test samples have a nominal gage section of 9 mm width by 12 mm thickness. We use commercially available Sylgard 184 from Dow Corning as the PDMS matrix material, a ~650 μm 3M hollow glass microsphere for the primary inclusion, and 27–32 μm Cospheric black polyethylene microspheres to construct a layer of fine tracking particles around the glass bead which serves as the DIC speckle pattern. The optical properties of all components allow for direct observation of the embedded DIC speckle pattern during the tensile test. This ensures that the calculation of the strain field surrounding the glass bead during loading and the debonding event can be achieved. A diagram of a fully constructed sample is shown in Fig. 1.

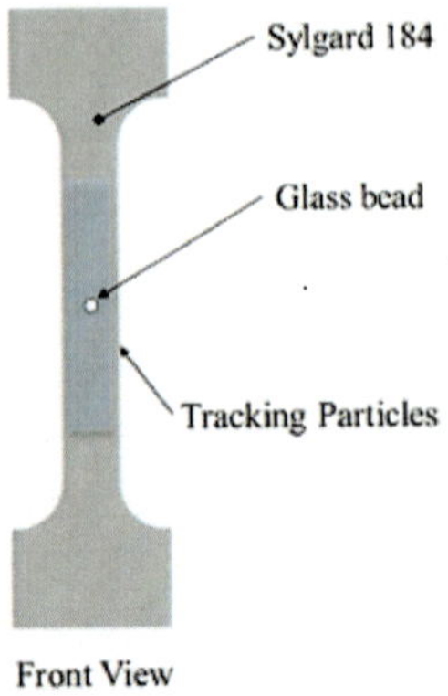

Fig. 1 Diagram of fully constructed tensile test sample with embedded DIC speckle pattern and glass bead inclusion [8]

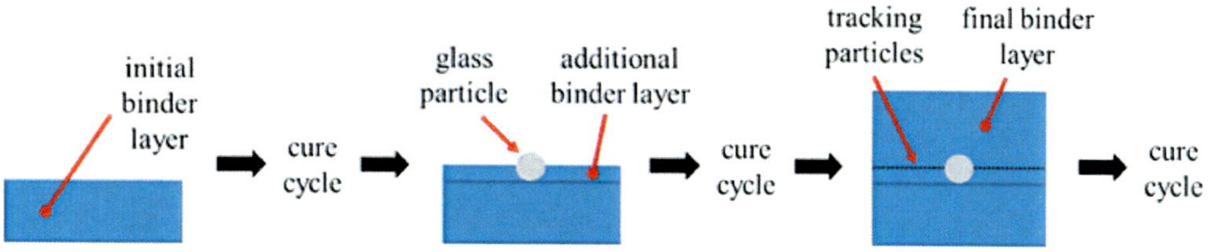

Fig. 2 Layered approach to constructing tensile test samples with embedded DIC speckle pattern and glass bead inclusion [8]

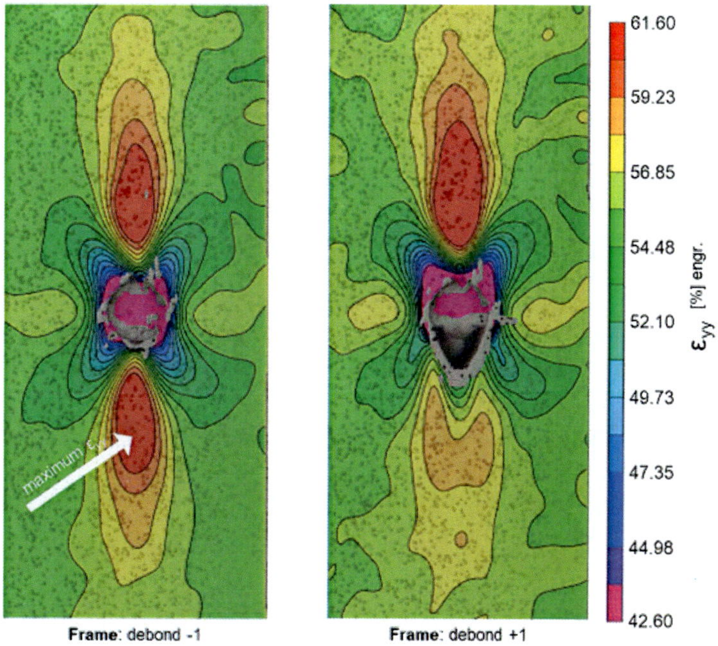

Fig. 3 The location of the maximum local tensile strain is indicated. Example of calculated strain field in the vicinity of the embedded particle in the vertical direction one frame pre-debond and one frame post-debond [8]

Construction of the samples is completed in a stepwise fashion with multiple casting and curing steps. This multiple step fabrication process enables placement of the glass bead at the approximate center of the dog bone and the DIC speckle pattern at the approximate mid-plane of the sample and bead. First, a base layer of matrix material is poured into the mold and then partially cured at 70 °C for 30 min to provide a solid base to place the glass bead upon. Second, a thin layer of matrix material is added to the partially cured base layer, the glass bead is placed, and the entire assembly undergoes a second cure cycle at 70 °C for 30 min. Third, the fine tracking particles are deposited on the cured surface of the second layer, and the remaining thickness of the mold is filled with matrix material. The entire sample assembly undergoes a third and final cure cycle at 70 °C for 30 min. The entire fabrication process is depicted in Fig. 2.

After fabrication, samples are stored under ambient conditions in the laboratory for a period of at least 3 weeks before testing. This time period is based on the results of an aging study we conducted recently [9] in which we found that the elastic modulus of Sylgard 184 continues to increase after fabrication for approximately 3–4 weeks and then reaches an asymptotic value. After the allotted time, each sample is tested using a conventional load frame. A tensile load is applied at 0.5 mm/s in order to apply a constant engineering strain rate. The evolution of the speckle pattern is captured by a 4 MP camera which is mounted on a translating stage to ensure the region of interest remains in the field of view of the camera due to the highly compliant nature of the elastomer matrix.

The resulting data includes a stress vs. strain curve from the load frame and a time series of images from the camera. The stress-strain curve provides the global stress in the gage section of the dog bone. The speckle pattern images are processed in a commercial DIC application to obtain the time-resolved two-dimensional strain field surrounding the glass bead inclusion from the unloaded state up to and through the debonding event. Figure 3 shows the calculated strain field in the vertical direction one frame pre-debond and one frame post-debond. The combination of the global stress measurement and the local strain field provides a rich data set describing the debonding process.

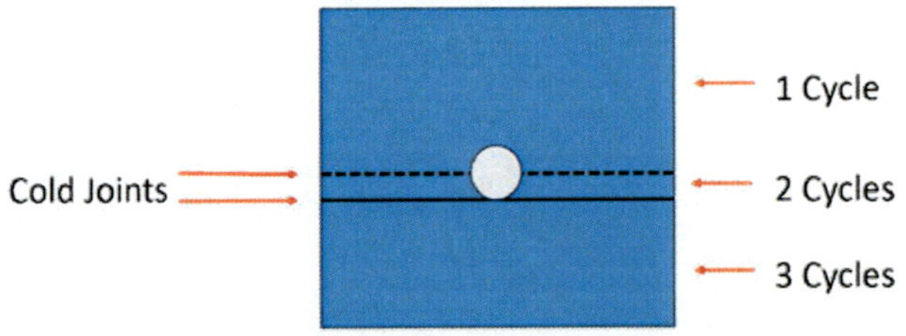

Fig. 4 Cross-section diagram of tensile dog bone sample highlighting the cold joints and functionally graded layers

Functional Grading and Cold Joints

While our fabrication approach is necessary for placement of the embedded glass bead and DIC speckle pattern, the presence of functional grading and cold joints caused by this process is unrepresentative of a homogeneous matrix material. Due to the multi-step casting and curing process, the base layer is exposed to three cure cycles, while the top layer is exposed to only one cure cycle. A fully constructed tensile sample consisted of three layers and two cold joints: the base layer, middle layer, and top layer with one cold joint between the base/middle layers and one cold joint between the middle/top layers, as depicted in Fig. 4.

Each cure cycle is prescribed at 70 °C for 30 min using a convection oven, with each casting step taking approximately 10 min. For reference, Dow specifies various cure times for Sylgard 184 ranging from 48 h at 25 °C to 10 min at 150 °C [10]. While each cure cycle is the same temperature and time duration, it follows that the base layer is cured for a total of 90 min, the middle layer is cured for a total of 60 min, and the top layer is cured for a total of 30 min. A natural consequence of this is that for a freshly fabricated fully constructed sample, each layer may have different mechanical properties. It is important to quantify this functional grading as it may have implications on data interpretation and computational model calibration. It is also important to quantify the effect of the presence of the cold joints for the same reasons.

To evaluate the effect of the functional grading and cold joints, we have devised two series of experiments to compare the effective bulk mechanical properties of homogeneous and multiple step constructed samples. The first series aims to specifically evaluate the functional grading, and the second series evaluates the presence of the cold joints. In both series, sample testing was conducted approximately 3 weeks after sample fabrication in accordance with the recommendations in our Sylgard 184 aging study to allow for the natural aging to stabilize [9].

For the functional grading study, three sets of six samples were produced. Each sample was cast to a "full" sample thickness of approximately 12 mm in a single casting step. Each set was then cured in a convection oven at 70 °C for 30, 60, and 90 min, respectively. The samples which were cured for 60 and 90 min were allowed to remain in the oven for the full duration rather than being pulled from the oven as they would be for a typical sample fabrication process. Each dog bone was individually packaged and stored in the same laboratory as the tensile load frame. All sample fabrication and testing were completed by the same operator using the same oven and load frame. Two different molds were used to expedite the fabrication process; however, both molds are nominally the same geometry. Tensile testing was completed using an MTS Exceed E42.503 5 kN load frame with a 1 kN load cell. All samples were aged for a prescribed 27 days.

For the cold joint study, three sets of 12 samples were produced. Each batch of dog bone samples was cast with one, two, or three layers to represent each stage of sample production. The first set was cast as a single layer of full 12 mm thickness with a single cure cycle for 30 min at 70 °C. The second set was fabricated in two steps, each 6 mm thickness, and two 30-min cure cycles at 70 °C. The third set was fabricated in three steps with each layer having the same thickness as a typical fully constructed sample and cured in three 30-min cycles at 70 °C. The first sample set of this series is essentially the same as the 30-min cure sample set of the functional grading series. Each dog bone sample was individually packaged and stored in the same laboratory as the tensile load frame. As with the functional grading series, all sample fabrication and testing were completed by the same operator using the same oven and load frame. Each set was fabricated using two available molds, with both molds having the same nominal geometry. Tensile testing was completed after samples were aged between 21 and 24 days using an MTS Exceed E42.503 5 kN load frame with a 1 kN load cell.

Results

For the functional grading study, batches of six samples were fabricated at three separate curing conditions for a total of 18 samples. Each batch was cured at 30, 60, and 90 min, respectively, at 70 °C and represents each of the three layers of a fully

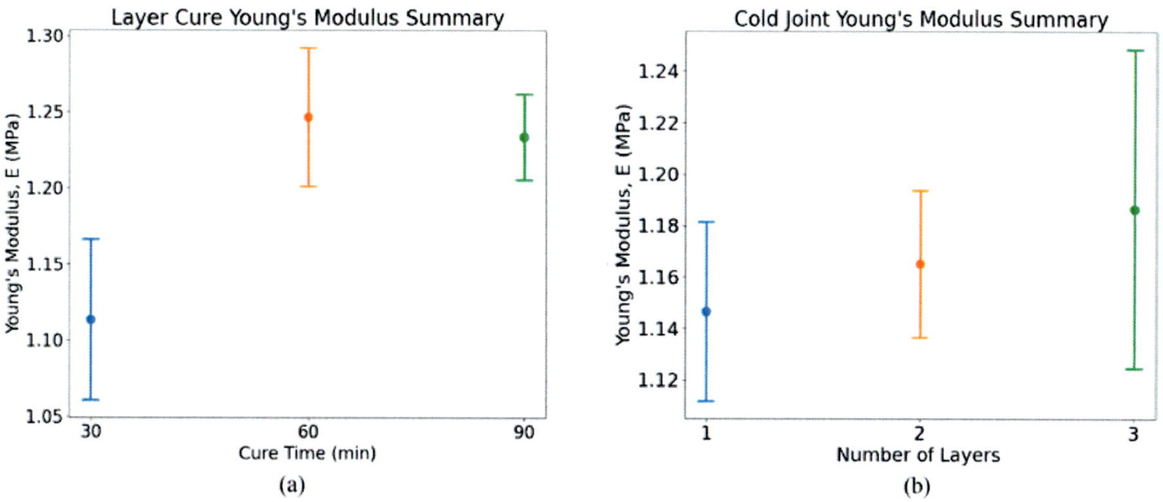

Fig. 5 Summary of results for functional grading study (**a**) and cold joint study (**b**)

constructed tensile sample (see right-most cross section of Figs. 2 and 4). The metric of interest from each tensile test was chosen as the slope of the stress-strain curve up to 25% strain. This slope represents the Young's modulus of the material and was determined using linear regression. The Young's modulus for the 30-, 60-, and 90-min cure samples was measured to be 1.114 ± 0.053 MPa, 1.247 ± 0.046 MPa, and 1.234 ± 0.028 MPa, respectively, where the values reported are the arithmetic mean and one standard deviation. These data are plotted in Fig. 5a. It is observed that the mean Young's modulus for the 60- and 90-min cure time samples is within one standard deviation between these batches, indicating no significant difference between the stiffness of the materials. However, it is observed that the mean Young's modulus of the 30-min cure time samples is 11.4% lower than the mean of the 60- and 90-min mean Young's modulus values. Additionally, the mean value of the 30-min cure time sample is outside of the standard deviation of the other two longer cured samples and indicates a more compliant material.

For the cold joint study, batches of 12 samples were fabricated for three separate layering conditions to produce a total of 36 samples. The first sample batch consisted of a single layer (no cold joints), the second sample set consisted of two layers (one cold joint), and the third sample set consisted of three layers (two cold joints). The metric of interest from each tensile test was again chosen as the slope of the stress-strain curve up to 25% strain to represent the Young's modulus of the material and was determined using linear regression. The Young's modulus for the single-, two-, and three-layer sample sets were measured to be 1.147 ± 0.035 MPa, 1.165 ± 0.029 MPa, and 1.187 ± 0.062, respectively, where the values reported are the arithmetic mean and one standard deviation. These data are plotted in Fig. 5b. It is observed that the mean of the moduli of the single- and double-layer samples lies within one standard deviation of the "full" three-layer samples. Furthermore, the difference of the mean Young's modulus between the three-layer system and the single homogeneous system is 0.04 MPa. These results indicate that there is no significant difference in the stiffness between the layered systems.

Conclusion

In this work we have built upon our previously developed technique [8] in order to further understand the capability and limitations of our approach. We have investigated the effect of functional grading and cold joints due to the sample fabrication process on the bulk elastic stiffness of our tensile test samples. We have conducted two test series to investigate these two effects and have concluded that there is no significant difference in the Young's modulus across each of the three layers. Additionally, we conclude that there is no significant influence of the cold joints on the bulk stiffness of our tensile samples. For further data processing, modeling, and simulation efforts, the stepwise constructed tensile dog bone samples can be considered as a simplified monolithic sample with minimal effects due to the functional grading and cold joints.

Acknowledgments The authors would like to thank the Air Force Office of Scientific Research for funding this work under grant 20RWCOR094. This work is Distribution A: Approved for public release, distribution unlimited (AFRL-2013-0813).

References

1. ASTM D1876-08(2015)e1: Standard Test Method for Peel Resistance of Adhesives (T-Peel Test). ASTM International, West Conshohocken (2015)
2. ASTM D4541-17: Standard Test Method for Pull-Off Strength of Coatings Using Portable Adhesion Testers. ASTM International, West Conshohocken (2017)
3. ASTM D6900-10(2015): Standard Test Method for Wet Adhesion of Latex Paints to a Gloss Alkyd Enamel Substrate. ASTM International, West Conshohocken (2015)
4. Tan, H., Liu, C., Huang, Y., Geubelle, P.H.: The cohesive law for the particle/matrix interfaces in high explosives. J. Mech. Phys. Solids. **53**(8), 1892–1917 (2005)
5. Gent, A., Park, B.: Failure processes in elastomers at or near a rigid spherical inclusion. J. Mater. Sci. **19**(6), 1947–1956 (1984)
6. Lauke, B.: Determination of adhesion strength between a coated particle and polymer matrix. Compos. Sci. Technol. **66**(16), 3153–3160 (2006)
7. Lauke, B.: Stress field calculation around a particle in elastic–plastic polymer matrix under multiaxial loading as basis for the determination of adhesion strength. Compos. Interf. **23**(1), 1–14 (2016)
8. Kosta, T., Mares, J.O.: Characterization of interface debonding behavior utilizing an embedded digital image correlation scheme. In: Lin, M.T., Furlong, C., Hwang, C.H. (eds.) Advancement of Optical Methods & Digital Image Correlation in Experimental Mechanics. Conference Proceedings of the Society for Experimental Mechanics Series. Springer, Cham (2021). https://doi.org/10.1007/978-3-030-59773-3_11
9. Kosta, T., Mares Jr., J.O., Hatter, M.M., Resue, B.M., Ridge, C.J.: The effect of aging on delamination strength utilizing an embedded digital image correlation scheme. In: Amirkhizi, A., Furmanski, J., Franck, C., Kasza, K., Forster, A., Estrada, J. (eds.) Challenges in Mechanics of Time-Dependent Materials & Mechanics of Biological Systems and Materials, Volume 2. SEM 2022. Conference Proceedings of the Society for Experimental Mechanics Series. Springer, Cham (2023). https://doi.org/10.1007/978-3-031-17457-5_13
10. "SYLGARD™ 184 Silicone Elastomer Technical Data Sheet." [Online]. Available: https://www.dow.com/content/dam/dcc/documents/en-us/productdatasheet/11/11-31/11-3184-sylgard-184-elastomer.pdf